Planning and Managing Agricultural and Ecological Experiments

Planning and Managing Agricultural and Ecological Experiments

Peter Johnstone

Routledge
Taylor & Francis Group
LONDON AND NEW YORK

First published 1998 by
Stanley Thornes (Publishers) Ltd

Published 2013 by Routledge
2 Park Square, Milton Park, Abingdon, Oxon OX14 4RN
711 Third Avenue, New York, NY, 10017, USA

First issued in hardback 2017

Routledge is an imprint of the Taylor & Francis Group, an informa business

A catalogue record for this book is available from the British Library.

ISBN 13: 978-1-138-40159-4 (hbk)
ISBN 13: 978-0-7487-3989-9 (pbk)

Contents

Preface

Experimentation is an essential feature of humanity's quest for knowledge, and a well-planned experiment gives generous rewards. It is surprising therefore to discover that many experimental programmes suffer from serious shortcomings. Out of 279 papers in 23 consecutive issues of the *Australian Veterinary Journal*, McCance (1995) found design weaknesses in 30%. Hurlbert (1984), drawing attention to 'some common ways in which experiments are misdesigned and statistics misapplied', says that 'Past reviews have been too polite and even apologetic'. The use of experimentation to improve industrial processes has been stimulated by Professor Taguchi, but his methods have been severely criticized by Box *et al.* (1988) who point out that Taguchi's methods 'are frequently statistically inefficient and cumbersome'. In an editorial in the *British Medical Journal*, Altman (1994) drew attention to 'The scandal of poor medical research' which included poor experimental design. Even statisticians have not escaped criticism. Ripley (1987) points out that 'Statisticians, even experts in the design of experiments, are notoriously bad at designing their own experiments!'.

This book has been written in an attempt to draw attention to the issues which need to be addressed in order to perform a satisfactory comparative experiment. The author has avoided presenting the mathematical and computational complexities of the traditional aspects of statistical design and analysis as modern computer programs facilitate their use without the need for such knowledge. Instead, references to the detail of the methods, which are admirably covered in many books, are given. It is hoped this non-standard approach results in an integrated presentation which maintains a strong focus on the planning and execution of an experiment without tempting the reader to become lost in the numerous theoretical and algebraic thickets along the way. The book emphasizes the important and often neglected issues of clearly defining the purpose of an experiment, the selection of experimental material, the measurements which need to be made, the modelling of responses, methods of comparison, power of comparisons, the use of covariates, experimental layout and the management of the conduct of the experiment and the data. Chapter 3 includes a brief and informal overview of the scope of some analytical methods. This is intended to provide the background for some planning strategies which are discussed later in the chapter and which result in efficient use of experimental resources. In Chapter 5, an account of classical statistical design strategies which involve replication, blocking and randomization is included. The sections concerning replication and blocking incorporate discussion of contemporary ideas of psuedo-replication and the use of blocks of natural size.

The book is intended for agricultural and ecological research students and researchers. It is also hoped that statisticians embarking on careers as biometricians will find the book useful. In writing the book the author has drawn on his 25 years of experience as a statistician working at an agricultural research centre specializing in soil, insect, animal and plant research. The level of statistical knowledge assumed is normally covered in introductory statistical courses and would include:

- descriptive statistics;
- probability distributions and random variables;
- basic concepts of hypothesis testing;
- basic ideas about inference from a population;
- comparison of treatments from a simple experiment;
- linear regression;
- randomized and factorial experimental designs.

Acknowledgements

Ever since books have been written, authors have prepared eloquent and grandiose acknowledgements expressing gratitude for assistance. I feel unable to compete in that formidable genre and so I simply but sincerely thank the following people who have helped me: my colleagues, Barbara Barratt, Neil Cox, Catherine Cameron, Barbara Dow, Dave Duganzich, Peter Fennessy, David Fletcher, Harold Henderson, Roger Littlejohn, Bryan Manly, David Miles, Alan Sinclair, Rhonda Sutherland, John Waller and Martin Upsdell for critical advice; Roger Littlejohn for making the GENSTAT procedure Appendix C more elegant; Nick Round-Turner for editorial comment; and Neil Cox for encouraging me to weed ruthlessly from the text any management jargon. I also make it clear that I accept sole responsibility for errors of fact and judgement.

Peter D. Johnstone

Experimental intentions 1

1.1 NATURE OF EXPERIMENTS

An experiment is a deliberate action or procedure, frequently referred to as applying a treatment, which is undertaken with the intention of provoking a response which the experimenter measures. However, the measured response is often difficult to interpret. This is because most experimental measurements are made up of two components. The first component is the real or true response to the applied treatment or combination of treatments. That is, it is the response if the treatment were precisely applied and the response was measured without error on experimental material which was absolutely representative of the **population** to which the results were to apply. The second component is **uncontrolled variation** and is the composite effect of the three potentially perturbing influences which are assumed to be absent from the true response. In a field experiment the uncontrolled variation is the composite effect of imprecision in applying the treatment, the effect of environmental factors such as uneven fertility, and the effect of imprecision in measurements. Similarly, in an animal experiment the uncontrolled variation is the variation in the individual's ability to respond, together with imprecision in applying the treatment and subsequent measurements. The uncontrolled variation masks the true response to the applied treatment in the same way as haze masks the true identity of landscape features. Consequently an observer is able to measure the overall effect, which is the sum of the two components, but is unable to measure the true response directly.

Experiments in which the magnitude of the real or true treatment responses are comparable to the magnitude of the uncontrolled variation are of particular importance and are the subject of this book. These experiments typically come from agriculture, ecology, medicine and industry. Statistical design techniques, when combined with statistical analytical techniques, enable an experimenter efficiently to separate and estimate the magnitude of the two components and so facilitate the drawing of defensible conclusions from the experimental measurements.

The classical and fundamental ideas of statistical design involve replicating the experimental treatments, allocating the treatments to the experimental material at random and minimizing the influence of uncontrolled variation. These will be referred to throughout the book, although Chapter 5 concentrates rather more on some specific issues concerning these aspects.

1.2 EXPERIMENTAL INTENTIONS

Before sensible experimental design and analysis strategies can be decided upon it is necessary to be specific about the purpose of a proposed investigation. One of the most eloquent pleas for this course of action was from Wilson (1974):

> There can be no greater waste of time and money than to embark on intricate programmes of applied research before one has a clear idea of the precise course of action one intends to embrace before, and I emphasise 'before', a single piece of data has been collected.

It is surprising how often this step is deftly avoided, although many words may be used to create the impression that it has been dealt with satisfactorily. It is important at the initial stages of an investigation to obtain a clear statement of intent, as many bewildering design questions are simply answered by referring back to these statements.

In order to amplify the idea consider the following example.

EXAMPLE 1.2.1

Many years ago a series of agricultural fertilizer experiments were stated to have had two intentions:

1. To delineate more closely the soil/climate/production conditions suitable for the effective use of a direct application of reactive phosphate rock (RPR) as a maintenance phosphate fertilizer.
2. To provide an initial screening of the agronomic performance of a variety of alternative RPR-based fertilizers relative to water-soluble fertilizers.

These non-specific intentions need to be translated into specific actions in order to undertake a focused investigation. The first step in this translation is to obtain a more specific statement of the intentions.

For example, it is necessary to define what is meant by 'the effective use of a direct application of reactive phosphate rock (RPR) as a maintenance phosphate fertiliser'. This will require making a decision concerning a measure of effectiveness. Is it to be the performance of a flock of sheep or herd of cattle, or some more direct measure of the productivity of a crop or a property of the soil? In either case, is the quality as well as the quantity to be compared? In order to say a fertilizer is effective its performance will need to be compared with a known standard. Consequently, it will be necessary to consider how the chosen measure of performance might be compared and the probability of detecting an important difference if it exists. For example, it might be decided to compare mean production at a specific application rate or, alternatively, the **parameters** of a response curve. Of course it is unlikely that the two fertilizers would produce exactly the same responses no matter how precisely they are measured. In view of this, how close are the measures of performance required to be before the two materials are to be considered to be equally effective?

As far as concerns the logistics of running the experiment, it will be necessary to consider what **variates** need to be collected, as the variates to be compared might not be the variates which can be directly observed. For example, if interest is in dry matter production over a specific interval of, say, four months, it will be necessary to cut and measure the production of the grown green herbage more frequently and this will need to be specified.

The 'soil/climate/production conditions' need close scrutiny. How many combinations are of interest? Is it to be an extreme and a mid-point of each combined in all combinations – that is, 27 combinations – or is it to be a random sample of all farms throughout the country, or a province? Are there specific combinations which are particularly important and only those combinations need to be investigated? The possibilities are infinite!

Perhaps the second requirement of the experiment, which is to provide an initial screening of the agronomic performance of a variety of alternative RPR-based fertilizers relative to water-soluble fertilizers, can be satisfied by listing the alternative candidates and asking if a subset would provide the required information. The chosen subset might be based, for example, on the solubility of the fertilizers in a standard solvent.

Generally, the requirement to translate the intention of a proposed scientific investigation into specific actions which need to be completed, soon leads to the need to answer important experimental design questions. These are:

1. What is the **experimental unit**?
2. What is a suitable selection of experimental sites?
3. What measurements have to be made?

4. How is the measured response to be described?
5. What comparisons need to be made?
6. What is the probability of detecting real differences between treatments? That is, what is the **power** of the comparisons?
7. What experimental layout is suitable?

Techniques which help experimenters answer these seven questions will be dealt with in subsequent chapters. However, it is only after the answers have been provided that experimental intentions can be translated into a work plan. This details specific tasks which need to be successfully completed in order to obtain the required experimental information. The preparation of a work plan is an essential part of the experimental process as it is necessary for assessing resource requirements and subsequent allocation. The exact methodology followed in this step does not matter so much as actually making a focused attempt at preparing such a plan. In fact, methodologies for translating intentions into a work plan are well practised in many fields of endeavour and many suggestions as to how it might be done are available to people requiring guidance. Basically, it involves translating an intention into a written document which details how and when specific tasks are to be undertaken and completed, and by whom.

1.3 JUGGLING INTENTIONS AND RESOURCES

In translating the intentions into specific actions, the physical and financial requirements for the conduct of the experiment soon become apparent. It is often found that the resources required are far in excess of those immediately available. Modifications to the intentions, and attempts to obtain more (rarely less) resources, are a time-consuming but essential part of the process. This iterative process of assessing the resources required for an experiment and modifying the intentions often goes through many cycles before the requirements of the experimenter and the available resources are matched. It is always possible they will never be matched. In this case it is worth considering the enterprising and courageous step of redirecting resources to or from other proposed experiments in order to do a single satisfactory experiment.

EXAMPLE 1.3.1

Suppose in Example 1.2.1 there is a requirement to reduce the resources required. In order to achieve this, the types of questions which may be asked include: is it possible to achieve the stated intentions with a subset of the available RPR-based fertilizers? Is it possible to limit the number of variates required to be observed without compromising the intention of the experiment? Is it possible to modify the intentions of the experiment and still find out something which is valuable? Is it possible to extend the time available for the experiment and reduce the size of each year's experimental programme? Is it possible to forgo the conducting of another experiment in order to satisfactorily complete this one?

1.4 FACILITATING THE PROCESS

The planning of complex experiments which involve a number of collaborators can be a time-consuming and contentious process. The purpose of a complex experiment is sometimes not clearly understood by all participants and, in addition, they usually have useful ideas they want to contribute. Some see the experiment as an opportunity to pursue a related but alternative investigation. Others simply misunderstand, sometimes due to a lack of fundamental knowledge. These ambiguities cause major difficulties which are becoming more important with present-day emphasis on multi-disciplinary, multi-site and multi-manager approaches to scientific endeavour. Many of the problems can be quickly resolved or at least ameliorated by a skilful facilitator who is knowledgeable in experimental design. The facilitator can be the project manager, a statistician or anyone else who has the necessary skills and is able to focus the attention of collaborators onto important issues and so speed the planning process to a successful and harmonious conclusion.

A facilitator also has a role in ensuring the subsequent management of the experiment is satisfactory. Meetings with facilitators can be very successful team-building experiences. As is argued in Chapter 6, this has a very positive effect on the quality of experimental data.

1.5 THE UNEXPECTED

Despite careful planning it is unlikely that an experiment will be conducted without something unexpected occurring. All experimenters hope the something unexpected will lead to an important discovery like Fleming's discovery of penicillin. However, it is far more likely that something will go seriously wrong. Hurlbert (1984) has called it 'demonic intrusion'. When faced with such occurrences it is unlikely that exorcism or human sacrifices will be as helpful as vigilance and a readiness to act sensibly and quickly in order to minimize the loss of data or the compromising of experimental intentions.

Selection of experimental material 2

2.1 IDENTIFICATION OF EXPERIMENTAL UNIT

Once the intentions of an experiment are clear, the experimental unit can be identified. It is the population entity about which some knowledge is sought. More formally, it is the entity about which hypotheses are to be formulated and inferences are to be made. It requires the experimenter to identify the population about which the results of the experiment are to apply.

It is important that each experimental unit has its allocated treatment applied independently of all the other experimental units. In this context 'independent' means that all the sources of uncontrolled variation are free to be fully expressed on each individual experimental unit. This ensures that the expression of uncontrolled variation is not conditional on what happened on any other experimental unit. As well as applying the treatment independently of the other experimental units it is necessary to ensure the response remains independent of the influence of treatments applied to other experimental units for the duration of the experiment.

There are two types of experimental units. The first is typically a single plant or animal. The second is typically a group of plants or animals. In order to distinguish between the two types of experimental units consider the following situations.

If inference is to be made about a single plant or a single animal, the experimental unit should be the single plant or single animal. For example, if inference is to be made about hormone dynamics within a plant or animal, individual treatments are independently applied to individual plants or animals. The individuals are then kept in such a way as to ensure the response to the treatment in any one of them is not influenced by the response in the others.

If inference is to be made about a field of plants or a group of animals then the experimental unit is the field of plants or the group of animals.

For example, if inference is to be made about yields of wheat cultivars when grown as a field crop, the experimental unit should be a field of wheat. The individual plants within the field do not behave independently of each other as they compete with each other for light and nutrients. If inference is to be made about growth of animals when maintained as a group then the experimental unit is the group of animals. Just as individual plants within a field do not behave independently, individual animals within the group do not behave independently of other members of the group. They interact socially with each other as they too compete for space and nutrients. However, it is important for the response in the experimental units, which are the field of plants or the group of animals, not to be influenced by the responses to treatments applied to other fields of plants or group of animals in the experiment.

It is common for the response to applying a treatment to a single plant or animal to be different than when the treatment is applied to a field of plants or a group of animals. This is because there is **interaction** between members of the group. The interaction will result in members of the group behaving in a different way than if they were on their own. A single field plant grown on its own will have a different habit than if grown in a group of similar plants. Similarly, if members of a group of animals are kept on their own they will not behave in a way typical of the way they would behave in a group.

In the above two examples, the field of wheat or the group of animals does not have to be as large as one might find in a commercial operation. However, it has to be large enough to behave in a way which is typical of the unit about which inference is to be made. In other words the experimental unit has to be typical of the system about which hypotheses are to be formulated and inferences are to be made.

Very often, although not exclusively, experiments on groups of plants or animals are concerned with the response to management on the treatment unit, and experiments on singletons are concerned with obtaining an understanding of the fundamental mechanisms which control and regulate the individuals.

EXAMPLE 2.1.1

It was required to determine the effect on reproduction in rabbits of the fungal endophytes which are found in the pasture grass tall fescue. This was to be done by feeding nine rabbits *ad libitum* pellets of regular feed and comparing them with nine rabbits fed pellets containing the high-endophyte tall fescue. The rabbits were to have a number of measurements made on them. The variates to be measured included feed intake, live weight, temperature, hormone levels, fertility and fecundity. The plan was to feed the rabbits their allocated feed pellets for two weeks. They were then to be mated and kept till they were due to give birth.

The purpose of the investigation was to estimate the size of the dose response of a certain diet in single rabbits. Inference was to be made about the response to this diet of reproductive performance. The individual rabbit was the experimental unit. As far as the experiment was concerned the best strategy for managing the animals would be to house them one rabbit to a cage. If this resulted in adverse effects on the rabbits, it would be universally applied to all animals. Consequently there would be no **bias** in favour of one of the treatments. The opportunity for group social interaction amongst the rabbits, if it was required, could easily be supplied without compromising the experiment. Provided animals were fed individually they could be given access, either individually or as a group of 18, to a recreation area.

The consequences of alternative housing arrangements were considered. As two cages, which would each hold nine rabbits, were already available, it was proposed to house the rabbits from each treatment group together in each of the cages. Unfortunately, this less expensive strategy would make the experimental unit a group of nine rabbits.

It seemed reasonable to suppose that rabbits, unaccustomed to living nine in a cage, would not behave in a way typical of single rabbits. Would there be unusual stress associated with the living arrangements that would have an effect on the variates being measured? An example of the type of thing which might occur would be that if one got an undetected infection, the chances of them all getting it would be high. This might lead to the temperature of all the animals in one cage being raised and being wrongly attributed. In addition, although it would be possible to measure the amount of food each group of nine animals ate, it would not be possible to know how much each individual had eaten. Consequently the opportunity to estimate a dose response curve of the supposed effect of the high-endophyte tall fescue would be lost. Additionally, with the group housing it is easy to envisage a situation where food intake of the individual rabbits would vary widely. Perhaps there would be some dominant rabbits present which would not allow others to feed, despite a generous supply being offered. If the hypothesis about the high-endophyte tall fescue was correct, this behaviour would lead to highly variable results. It was suggested that the individual food intake could be estimated indirectly from each rabbit's weight change during the course of the experiment. However, weight changes are not a direct measure of the size of effects due to changes in intake and so would be a very poor substitute for the measurement they were being proposed to replace.

The uncertainties resulting from treating the animals as a group when interest was in individuals lead to serious doubts about interpreting the proposed experiment. After reviewing the consequences of using group housing, the experimenters decided to forgo the cost saving such a strategy offered, and perform a more expensive experiment with one rabbit per cage. They preferred an experiment in which the experimental unit was correctly identified as it had a better chance of providing an unambiguous interpretation.

EXAMPLE 2.1.2

Sheep are usually grazed together and any experiment concerned with the management of flocks of sheep should have as the experimental unit flocks of sheep which are large enough to be considered to behave in a typical way. Consequently, if it is required to know something about the potential of certain pasture species as a stock food for groups of lambs, it is necessary to carry out the investigation on groups of lambs which are supposed to behave in a typical way.

In an investigation into the growth of lambs grazing various pasture species on a number of farms over a number of years, the components of uncontrolled experimental variation expressed in terms of the different components of **variance** have been estimated as follows:

- Between sites 50 kg^2
- Between flocks (in different years), within sites 190 kg^2
- Between lambs within flocks 25 kg^2

It is clear from these estimates of **variance components** that the variation between lambs within flocks is much less than the variation between flocks within sites. That is, the response to applying a grazing treatment to a flock of lambs is likely to be far more variable than the variation in the response of individual lambs within any particular flock. This has serious consequences when recommendations for grazing management are given after observing the response to that treatment on just one flock in one year.

2.2 SELECTION OF EXPERIMENTAL SITES

Two main types of experiment are distinguished. The first type of experiment uses a single environmental condition, often controlled, and is normally designed to help understand the fundamental mechanisms of the system being studied. Results are generally applicable as they do not vary with varying conditions. The second type of experiment is designed to measure the variation in response of the system being studied when it varies under varying conditions.

The selection of sites for the first type of experiment should present no problems. The experimenter uses an available facility and gets on with the job. When interpreting the results from this first type of experiment the possibility that the observed response may be different in different environments needs to be considered.

The second type of experiment poses many real problems. A random selection from a population of sites about which conclusions are to be drawn allows an unbiased estimate of the overall comparisons of interest and an estimate of its variability. This is the survey approach and is used infrequently because of the difficulties of first choosing the sites and then gaining access to them. However, random selection of sites should not be discarded without due consideration of the consequences. As Jeffers (1992) says: 'No amount of modelling and computation can compensate for the failure to ensure that data are obtained from objective samples of carefully defined populations about which inferences are to be made.'

Unfortunately, the more usual approach to the second type of experiment is to make a non-random selection of sites. This allows comparisons to be made on environments which are thought to be important, often called representative, and conclusions are generalized to other sites on the basis of what is supposed to be known about those sites. The selection of sites can be made for the wrong reasons, such as easy access, a co-operative or influential collaborator or the proximity of servicing staff. Seriously biased results can result from these practices.

EXAMPLE 2.2.1

In Example 1.2.1 it was stated that the intention was to delineate more closely the soil/climate/production conditions suitable for the effective use of direct application RPR as a maintenance phosphate fertilizer.

This implies that a range of sites with specific selected conditions would be chosen for the conduct of the experiment. This was in fact the case. Nineteen sites were chosen and these were selected on fertility status, history of fertilizer application and relevance to important farming areas. This selection implies a random sample of farms from important areas would have been a useful design strategy. However, it was decided to forgo random selection in favour of choosing sites farmed by co-operative owners near servicing centres. The cost of the trade-off between convenience and good scientific practice will never be known. However, the risk that the intention of the experiment would be compromised is a risk many researchers would not be prepared to take.

2.3 MEASUREMENTS TO BE MADE

In many investigations the responses of interest and the variates which are directly measured coincide, but this is not necessarily so. Many responses which are of primary interest cannot be measured directly and are calculated as sums, differences, products, ratios or combinations of these operations from the variates which are actually measured. An example is the dry matter yield from a field, expressed as a yield per hectare. It is calculated from the weight of green material harvested from a small area of the field along with the dry matter content which is determined from a sub-sample of the harvested green material.

Other frequently used responses calculated from measured variates are from time sequences of measurements on the same experimental unit. Interest in these types of data often centres on changes or a maximum over a specified time interval, the area under the curve, a measure of cyclic behaviour such as periodicity, the estimated parameters of a curve which characterizes an interesting aspect of the changes over time, or some other summary **statistic**.

Less frequently, new variates are calculated from a linear combination of many variates measured at the same time. An example is the calculation of a new variate by combining the counts of various larval and adult stages of a number of species of parasites occurring within an animal. Such a variate may be interpreted as an index of parasitism. **Principal components** analysis is a technique used to create new variates from linear combinations of other variates. The principal components are independent linear combinations of members of a set of variates. The first principal component has the maximum possible variation of any linear combination of observed variates, and the second and subsequent principal components have the maximum possible variation while remaining independent from the previous principal components.

A similar idea is that of **canonical variates**. Canonical variates are pairs of independent linear combinations, one from each of two sets of variates. The pairs of linear combinations are such that the **correlation** between each pair is the maximum possible. If treatments applied to different experimental units are indexed, from 1 to the number of treatments, and linear combinations of these indices are used as one of the sets of canonical variates it is possible to use the technique of canonical variates to obtain a rule which assigns individuals to one group or another. This technique is often referred to as **discriminant analysis**. A famous example of the use of this technique is given by Fisher (1936). The sepal length and width and petal length and width were measured on each of 50 flowers from each of three varieties of iris, namely *Iris setosa*, *I. versicolor* and *I. virginica*. When discrimination analysis was performed on the data from *I. versicolor* and *I. setosa* it was found that the following linear combination of the four variates,

$$\text{sepal length} + 5.9037 \times \text{sepal width} - 7.1299 \times \text{petal length}$$
$$- 10.1036 \times \text{petal length},$$

gives a mean value for *I. versicolor* of 66.917, and for *I. setosa* of –38.424. The mid-point between the mean values for the two species is 14.247. Consequently the value of this linear combination of the four variates may be used to discriminate between species, and measurements made on an unknown flower could be combined in this formula and used to identify *I. versicolor* if the value was greater than 14.247 and *I. setosa* if it was less than 14.247. The standard deviation of the new variate, when calculated from the within-species variation, enables the calculation of the probability of misclassification. In general, discriminant functions are used when any one variate is not sufficient to make a certain allocation of an individual to a group. See Manly (1986) for an accessible and detailed account of these methods.

The possibilities for using calculated variates are endless. One example concerns the distance from a random point to the closest specific plant and the distance from the closest specified plant to the next closest. These measurements can be used to estimate plant density and, if the plant is harvested, yield per unit area (Diggle 1975). In some situations this technique might have considerable advantages over mowing all the herbage in a plot and then dissecting out the individual species.

Measures of variability are useful in some management or technique comparison studies. An example is the estimated within-group standard deviation of lamb weights. A uniform group of animals is usually more valuable than a heterogeneous group of the same average weight. Another example is the standard deviation of product from alternative suppliers. A more uniform product is usually more desirable.

It is wise to exercise restraint when collecting data. The collection of data just because they are available and because they may eventually prove to be useful is, in most situations, extravagant. The urge to collect data which are not required for the stated experimental intentions must be resisted in much the same way as scavenging must be resisted when visiting bargain basements or community refuse sites. In both situations a very attractive item often proves to be not as useful as first thought, or, if it is useful, it is soon discovered that some essential component is missing.

Care is needed to ensure the variate which is to be modelled can be interpreted in the way it is required to be interpreted. For example, a pasture harvested by mowing at frequent intervals does not necessarily yield the same as if it were harvested continuously by grazing animals. Differences are due not only to the different frequency of harvesting, but also because of wrenching of the plants, treading, urinating and defecating by animals and selective grazing of preferred species. It would be a mistake to make a direct translation of information from a mowing experiment to grazing animals without the support of supplementary information. I know of one situation where 15 years of careful selection of a pasture species under a mowing regime was demonstrated to be of no value at the first grazing by farm animals. The carefully nurtured plants were all uprooted because of their insecure root habit. Another example concerns the interpretation of the presence of potentially lethal virus particles in dead insect larvae. It was hoped to interpret the presence of a virus in a dead larva as the cause of death but this does not necessarily follow. Dead victims of road accidents will contain many potentially pathogenic organisms but it certainly is not the organisms which are responsible for their host's death.

Describing or modelling responses 3

3.1 LOGIC OF SCIENCE

In order to make sensible decisions about how to describe and interpret measured experimental responses taking into account current accepted theories and other reported observations, it is important to review briefly the logic of science. Karl Popper (1959) in *The Logic of Scientific Discovery* started a revolution in the way many scientists think. His view is that it is not possible to prove the truth of a scientific theory; it is only possible to falsify it.

The old logic of scientific thought, based on a search for natural laws, was described by Francis Bacon. The bare bones of the method are as follows. Scientists would make measurements on a phenomenon for which an understanding was being sought. As the number of measurements increased and time passed, a tentative law was formulated which explained all the known facts. Much energy was now expended on verifying the law by finding supporting evidence. If this was successful a new law had been found and the store of scientific knowledge had been added to.

This inductive approach was used into the twentieth century despite the philosopher David Hume having raised some awkward questions in the eighteenth century, which had been conveniently overlooked. Hume pointed out that just because something had always happened there was no guarantee it would always happen in the future. Because of the psychological way we are constituted we might expect something which had always happened to happen again. However, there is no logical reason why it should.

Popper has been able to present a solution to Hume's problem. It requires the replacement of the traditional philosophy of science with another. It is beautifully simple. It says that in logic a scientific law is conclusively falsifiable but it is not conclusively verifiable. Even so, there are severe methodological problems with this simplicity. It is always possible to find a reason to doubt a measurement and so to discount falsifying evidence. Also, in order to facilitate the adoption of new theories, Popper advocated that they be stated unambiguously. It is then possible to design an experiment which would give falsifying evidence. Nevertheless, there has to be a balance whereby current theories are not abandoned lightly.

New theories are created by individuals exercising their imagination on the disparity between measurements and the predictions from existing theory. They are human inventions and it is not possible to know if they are correct. Sometimes new theories result from reassessing already known facts or new non-experimental information. Typical examples come from cosmology and geology. However, comparative experiments, if they can be performed, can be expected to result in more rapid progress.

Scientific knowledge is best thought of as a set of conjectures. This is illustrated by Newton's and Einstein's differing views of the universe. Newton invented his theory and at the time it fitted all the known facts. It continues to this day to provide a satisfactory model of many physical events. However, as time passed some anomalies were observed between what happened and what was predicted to happen if the theory were correct. Einstein then invented his theory which explained some of these anomalies. Despite this, Einstein thought his theory was still defective and spent the second half of his life trying to find a better one.

It is quite admissible to have two theories, both of which are accepted at the same time. An example is the modelling of light as both particles and waves. As long as a particular theory does what is required of it in predicting the future and explaining observed phenomena, it is perfectly adequate for that purpose. Despite this, however, it must be recognized that a theory which is suitable for one purpose might be totally unsuitable for another.

3.2 DESCRIBING OR MODELLING RESPONSES

An acceptable theory concerning a natural phenomenon is expressed as a model. This can be used as a succinct statement of the theory and as a means of predicting the future. Modelling is an essential part of everyone's life and there should be no mystique about it. Models may be expressed in mathematical ways, but this is not necessarily so. Everyone knows that a ball thrown into the air will come down more quickly than a feather. This is an acceptable model and most people require nothing more. However, it is also possible to model the passage of the ball and feather through the air in a very sophisticated mathematical way, taking into account all the factors which are known to influence the process. Clearly, the adequacy of a prediction from a model and its sophistication depend solely on for what it is required.

An important part of designing an experiment is to select possible alternative models for the experimental measurements. This should be done before the commencement of the experiment. Model selection must realistically reflect the purpose of the experiment. Once possible models have been selected the **precision** of the predictions and the ability to discriminate between the alternate models and theories are important considerations which have to be addressed. It is only after this has been done that the resources required for the conduct of a satisfactory experiment can then be identified. If it is not done, it is a matter of luck as to whether or not the experiment is adequate for the intended purpose. So as to facilitate model selection some frequently used models are discussed below. Ways to improve precision of predictions from these models and others by careful experimental design strategies are discussed in sections 3.3 and 3.4 and in subsequent chapters.

A most important description or model is the additive model. The aim of this model is to describe responses to different treatments in the simplest way. The model includes a term which describes the real effect of a treatment plus a term which describes the uncontrolled variation associated with the measurement. It is an algebraic restatement of the formulation of the nature of experiments presented in section 1.1. A simple additive model, where a measured response Y is attributed to the action of a specific treatment plus uncontrolled variation is

$$Y = \tau_i + \varepsilon,$$

where τ_i is the additive real or true response to the ith treatment and ε the uncontrolled variation in the measured response. The simplest model used in comparative experiments describes the responses to just two treatments.

EXAMPLE 3.2.1

Suppose it is required to describe the response to two animal feeds, regular pasture and chicory, and the measured response is to be the average copper levels in the livers of flocks of sheep grazing the fodder. The experiment is replicated on four different farms. If the real or true response as defined in section 1.1 was 2500 micromoles of copper per gram and 3600 micromoles of copper per gram, respectively, the measured response may be described by the model $Y = \tau_i + \varepsilon$ for $i = 1$ and 2. Possible measured responses and their two components might be as follows:

Farm number	Average liver copper levels for flocks grazing regular pasture, i.e. $Y = \tau_1 + \varepsilon$	Average liver copper levels for flocks grazing chicory, i.e. $Y = \tau_2 + \varepsilon$
1	1690 = 2500 − 810	3110 = 3600 − 490
2	1900 = 2500 − 600	3190 = 3600 − 410
3	2680 = 2500 + 180	3460 = 3600 − 140
4	3540 = 2500 + 1040	4670 = 3600 + 1070

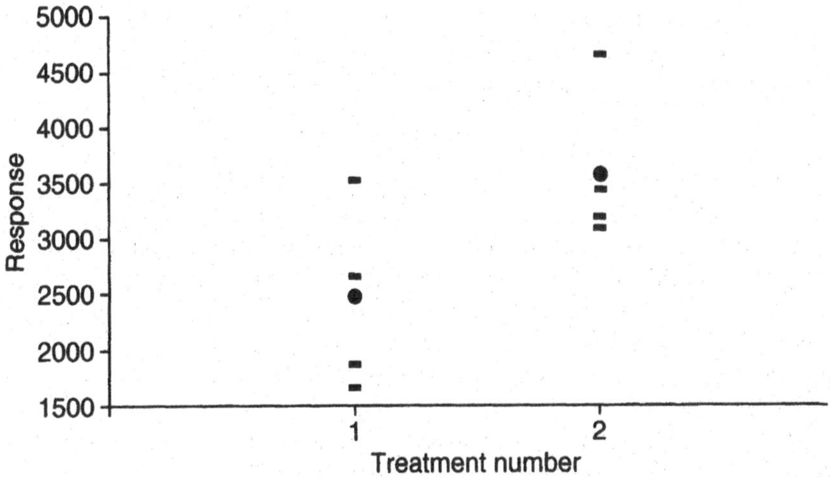

Figure 3.2.1 Real response (dots) and measured response (dashes) from Example 3.2.1.

The same data may be displayed graphically as shown in Figure 3.2.1. The dots represent the real or true responses to treatments 1 and 2 and the dashes represent the measured responses on the four farms.

The τ_i of the simple additive model can be thought of as either unstructured or structured. An example of unstructured treatments is presented above in Example 3.2.1. They are unstructured as the two treatments are not related to each other in any quantitative way. Another example of unstructured treatments is in the model which describes the responses of many varieties of the same species of a plant. Here, the many varieties are conveniently classified into a single group or set of related treatments known as a **factor**. Each different variety within the group or set constitutes a level of the factor and consequently the number of factor levels equals the number of varieties and the levels are used to index the varieties.

In contrast to unstructured treatments, structured treatments are those where the levels of the treatment factor are related to each other in a known and meaningful way. An example is when fertilizers are applied at the rates of say 200 kg/ha, 100 kg/ha, 75 kg/ha, 50 kg/ha and a control of 0 kg/ha. In this example the treatment factor is fertilizer and it has five levels which differ from each other in a known and quantifiable way.

In this situation, instead of modelling the response by

$$Y = \tau_i + \varepsilon,$$

where the τ_i are the five different responses to the five different rates of fertilizer, it may be more informative to model the response by a curve. For example, measurements may be modelled by

$$Y = \alpha + \beta \rho X + \varepsilon,$$

where α is the maximum true response, β is the difference between the true responses when no fertilizer is applied and α . The parameter ρ is restricted to $0 < \rho < 1$ and it controls the rate at which increasing applications of fertilizer – that is, X – increases the true response. Such a curve is often called the curve of diminishing returns because as X increases, the magnitude of the response for incremental increases in X decreases. Figure 3.2.2 demonstrates the shape of the curve where α = 11 000, β = –2500, ρ = 0.98 and ε = 0.

Measurements from an experiment would include a component of uncontrolled variation – that is, the ε of the above model. Consequently experimental measurements would be scattered about this true line as they were about the true responses in Example 3.2.1.

Other models are commonly used to describe experimental measurements. For example, a response may be modelled by a straight line

$$Y = \alpha + \beta X + \varepsilon,$$

where α and β are the intercept and slope, respectively, and the response is dependent on X, a predictor variate. Experimental measurements would be scattered about the true line given by $\alpha + \beta X$ because of the uncontrolled variation in the measurements. An example is presented in Figure 3.2.3, where it is known that a true response can be described by the equation $Y = 5 + 10X$ with X taking the values between 1 and 10. Measured responses obtained from an experiment designed to estimate the relationship will include a component of uncontrolled variation and are unlikely to take the true values 15, 25 and so on to 105.

The idea of having a model with just one factor with levels τ_i may be extended. A factorial model is one where two or more factors are combined in every combination of levels. By convention a factorial

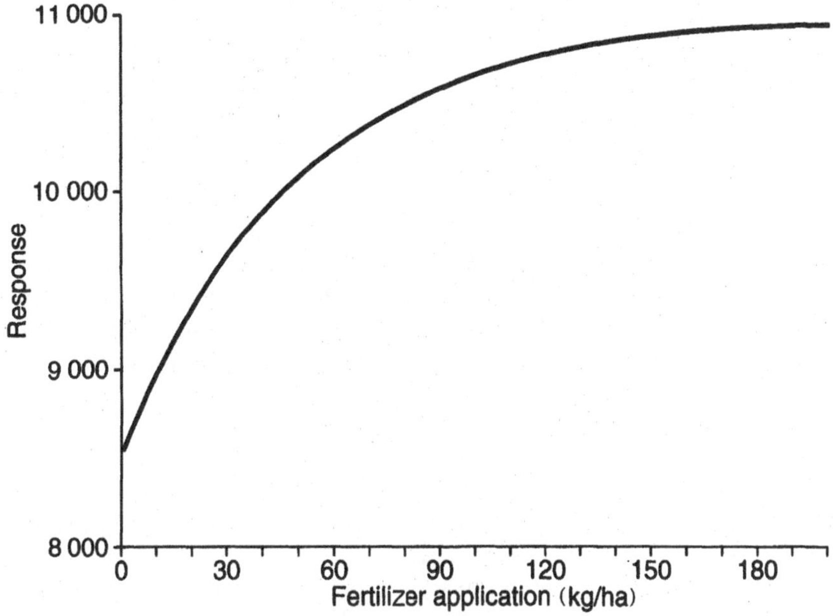

Figure 3.2.2 Real response to fertilizer application.

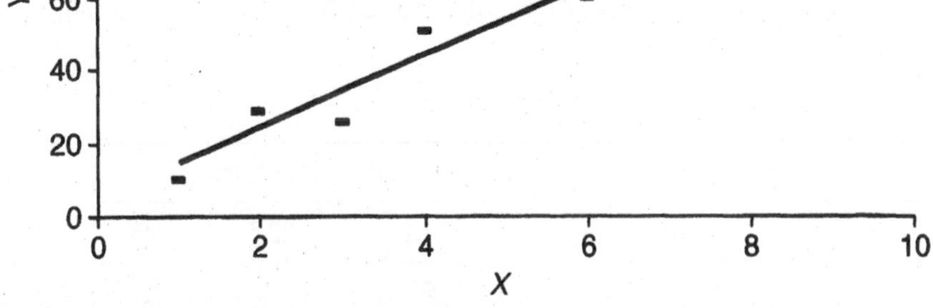

Figure 3.2.3 Real response (line) and measured response (dashes) for the equation $Y = 5 + 10X$.

model is known by the product of the number of levels of each factor. For example, if one factor has five levels and another has four levels, the model is a 5 × 4 factorial model. Factorial models enable the modelling of the different responses to a factor at each of the levels of another factor. That is, they enable the modelling of the interaction between the factors. A model involving two factors is frequently expressed as

$$Y = \varphi_i + \kappa_j + \gamma_{ij} + \varepsilon,$$

where φ_i and κ_j are the overall response to factors P and K at levels *i* and *j*, respectively, and γ_{ij} is the interaction which is the increment of response which is unique to the *ij*th combination of factor levels. There are alternative ways of modelling such a response, as is shown in the following example.

EXAMPLE 3.2.2

Suppose it is required to model the response to increasing application rates of sulphur fertilizer while increasing the application rate of phosphate fertilizer. It is assumed there are good reasons to have five rates of sulphur application and four rates of phosphate. All the possible factor combinations, 20 in all, are described as a 5 × 4 factorial experiment. The 20 treatment combinations are presented in the table below.

This would enable the modelling of the changing response to sulphur with increasing applications of phosphorus. This might be done by allowing a different curve for the sulphur response for each level of phosphorus. However, there are alternative ways of modelling the interaction, and the one chosen would depend on the stated experimental intentions and would suit that particular requirement. One possible way of modelling the response would be as the surface displayed in Figure 3.2.4. This may be used to determine where the system is potentially unstable; such as when small changes in one fertilizer have a large influence on the response. In the surface of Figure 3.2.4 this is the area in the neighbourhood of zero application for both phosphorus and sulphur.

		Sulphur (S) application				
	kg/ha	0	11.25	22.5	45	90
Phosphorus (P) application	0	P_0S_0	$P_0S_{11.25}$	$P_0S_{22.5}$	P_0S_{45}	P_0S_{90}
	40	$P_{40}S_0$	$P_{40}S_{11.25}$	$P_{40}S_{22.5}$	$P_{40}S_{45}$	$P_{40}S_{90}$
	80	$P_{80}S_0$	$P_{80}S_{11.25}$	$P_{80}S_{22.5}$	$P_{80}S_{45}$	$P_{80}S_{90}$
	160	$P_{160}S_0$	$P_{160}S_{11.25}$	$P_{160}S_{22.5}$	$P_{160}S_{45}$	$P_{160}S_{90}$

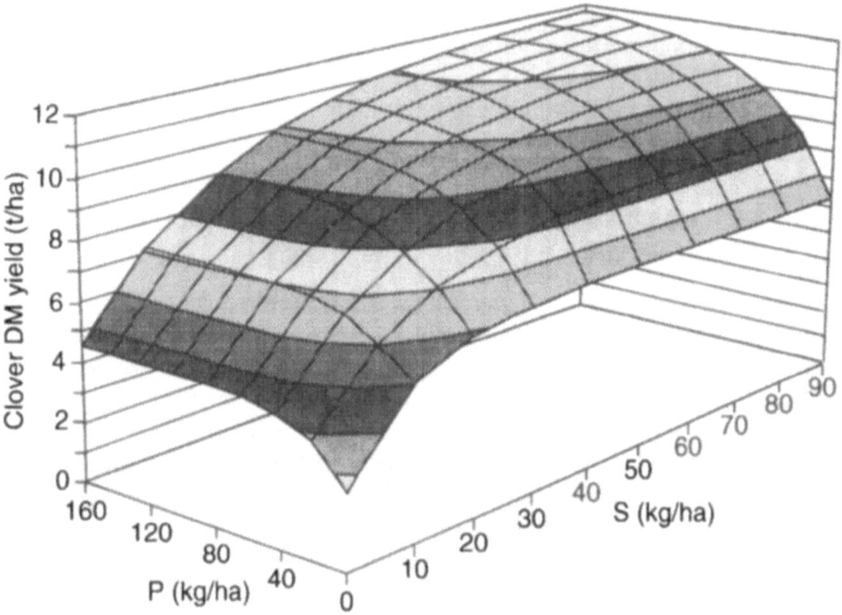

Figure 3.2.4 Response to sulphur and phosphate applications modelled by a surface.

Another use would be to determine when an expensive nutrient was being wasted because of deficiencies in a cheaper one. In Figure 3.2.4 this would be when phosphorus was applied with low levels of application of sulphur.

When factorial models have more than two factors the number of individual combinations of treatments increases rapidly. In order to illustrate the rate at which this escalation occurs, consider the following technique which is used to determine which nutrients are limiting plant growth. It involves applying an adequate amount of each nutrient and comparing the response when no nutrient is applied. That is, each nutrient is a factor with two levels, one level being zero application and the second being what is judged to be an adequate amount. If eight nutrients are being investigated there are $2^8 = 256$ individual treatment combinations. In order fully to describe the response to such an experiment a very complex model is required which involves interactions between all combinations of factors. In such situations, it is possible to reduce the number of model terms by excluding interaction terms involving many factors. It is common practice, although by no means a general rule, to exclude interaction terms involving more than three factors. Such an action is justified when prior knowledge indicates that such interactions are likely to be small and unimportant. This strategy of limiting the number of interaction terms in the model has an important consequence on the design of factorial experiments which will be discussed in section 3.3.

3.3 ESTIMATING MODEL PARAMETERS FROM EXPERIMENTAL MEASUREMENTS

So far, when models have been discussed, it has been assumed that their parameters are known. In Example 3.2.1 it was assumed we knew what the real or true values of the average liver copper levels were. In Figures 3.2.2, 3.2.3 and 3.2.4 it was assumed that the real or true values of the equations from which the figures were drawn were known. The assumption that they were known was made in order to demonstrate the idea that an experimental measurement is composed of two components. In any real experiment it is impossible to know any of these real or true values and the whole purpose of the experiment is to make measurements from which estimates of the unknown parameters of the models are obtained.

Many techniques are available to estimate model parameters from experimental measurements. The three which will be discussed are least squares, maximum likelihood and residual maximum likelihood. These methods are objective ways of fitting a model to data and usually they obtain a result which is similar to that obtained when the data are fitted by eye. This makes the methods intuitively plausible. They will now be described informally in order to make some points relating to the designing of experiments.

3.3.1 LEAST SQUARES

Least squares is the most common technique used to estimate model parameters. The treatment responses in the additive effects model of section 3.2 are usually estimated by least squares. In that form it is often called the analysis of variance (ANOVA). In this situation the least-squares estimate of a treatment response is the mean of the measured responses on the experimental units to which that particular treatment was applied. The fitted model which is used to estimate the true response to the *i*th treatment and hence predict future responses is expressed as follows:

$$\hat{Y} = \hat{\tau}_i.$$

The circumflex (ˆ) above the Y indicates that the response is an estimated value as distinct from an actual measurement. Similarly, $\hat{\tau}_i$ indicates it is an estimate of the parameter τ_i rather than the true value. The mean of the average liver copper levels for each of the four treatment flocks of Example 3.2.1 is 2452.5 micromoles of copper per gram for those grazing regular pasture and 3607.5 micromoles of copper per gram for those grazing the chicory pasture. It should be noted that these estimates are not the same as the true effects which were 2500 micromoles of copper per gram and 3600 micromoles of copper per gram, respectively.

The form of the model fitted by least squares may be a straight line where the estimated true responses, $\hat{Y}$, are given by the equation

$$\hat{Y} = \hat{\alpha} + \hat{\beta} X,$$

where the circumflexes above the α and β indicate they have been estimated from experimental observations. The estimates, $\hat{\alpha}$ and $\hat{\beta}$, are used to obtain a prediction of future responses $\hat{Y}$ for given values of X.

The relationship which is used to model a response may be a surface which looks like a sheet of suspended plywood. If this is the case, the estimated response which is used for predicting future responses is given by the equation

$$\hat{Y} = \hat{\alpha} + \hat{\beta} X + \hat{\gamma} Z,$$

where $\hat{\alpha}$, $\hat{\beta}$ and $\hat{\gamma}$ are the least-squares estimates of the unknown parameters from experimental measurements, and X and Z are predictor **variables**.

The equations $\hat{Y} = \hat{\tau}_i$, $\hat{Y} = \hat{\alpha} + \hat{\beta} X$ and $\hat{Y} = \hat{\alpha} + \hat{\beta} X + \hat{\gamma} Z$ are linear equations as the predicted responses are calculated as linear combinations of the parameter estimates. Least squares can be used to estimate parameters of equations which are not linear. An example is the asymptotic or exponential curve, where the estimated response and predictions are given by

$$\hat{Y} = \hat{\alpha} + \hat{\beta} \hat{\rho}^{X},$$

in which $\hat{\alpha}$, $\hat{\beta}$ and $\hat{\rho}$, $0 < \hat{\rho} < 1$, are estimates of unknown parameters α, β and ρ, and X is a predictor variable.

When the response to be modelled is a continuous function as in the last three examples, least squares is often called regression analysis.

The idea of least squares is demonstrated in Figure 3.3.1. Values of the unknown parameters of the model which is to be fitted are chosen to have particular values which are substituted into the equation to give the $\hat{Y}$s, the estimated or predicted responses. The parameter values are chosen such that the sum of the squared distances between the measured experimental values, the Ys, and the $\hat{Y}$s, is minimized. The sum of the squared distances, the sum which is minimized, is the sum of the squared vertical distances between the squares and dashes shown in Figure 3.3.1. This can be found by trial and error, although analytical methods are usually

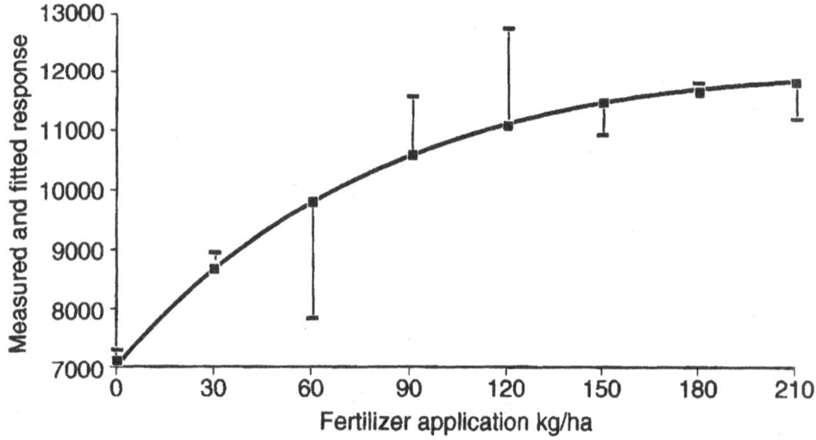

Figure 3.3.1 Fitted response (squares) and measured response (dashes) to fertilizer application.

far more efficient. There are many computer implementations of these analytical methods and they are available in most programs which claim a statistical analysis capability.

However, before using a program it is important to obtain guidance on its proper use as each has its own unique syntax. In addition, the generality of the programs varies enormously and some have a very limited capability. It is important to understand a chosen program's limitations in order to avoid frustrations in attempting to coax from it an analysis which is beyond its capability. A frequent limitation is the inability to model blocks, the use of which is discussed in section 5.2. SAS (SAS Institute, Rayleigh, North Carolina, USA) and GENSTAT (Rothamsted Experimental Station, Herts, UK) are amongst the most general programs, but for a non-expert it is important to obtain guidance in their use. Such programs are usually the most difficult to master, and the use of templates, set up by an experienced user, allows a novice a degree of independence with only a small amount of instruction on how to modify the code.

If the model adequately describes the measurements, and they are representative of all possible measurements, the squared distances between the Ys, and the $\hat{Y}$s, provide an estimate of the variance of the uncontrolled experimental variation. In example 3.2.1 this is 615 592. Fisher (1925) recognized that random allocation of the treatments to the experimental units ensured that measurements possessed this desirable property. Therefore, it is fundamental and important to allocate treatments randomly to experimental units. This requirement is one of the foundation stones of modern experimental design and is dealt with more fully in section 5.3.

Least squares provides parameter estimates which are unbiased provided they have come from a randomized experiment and are the parameters of a linear function. The term 'unbiased' means that there is no systematic distortion in the estimated parameters from a large number of estimations from different data sets. With non-linear functions the estimated parameters are not necessarily unbiased, but in most situations the biases are not large enough to compromise the purpose for which they are required, although that possibility needs to be borne in mind.

As well as providing parameter estimates, least squares provides estimates of uncertainty in the parameter estimates. These are called the **standard errors** of the parameter estimates. In Example 3.2.1 the standard errors of the parameter estimates, the means, are 392.3, and of their difference, 554.8. The uncertainty is a consequence of the uncontrolled variation in the measurements from which the parameter estimates are calculated. The size of the standard errors is directly dependent on the size of the variance of the uncontrolled variation, an estimate of which is obtained with the least-squares estimates of the parameters. Consequently, any successful attempt to minimize the uncontrolled variation will reduce the uncertainty in the parameter estimates and subsequent predictions of responses. Methods which facilitate this objective are discussed in sections 4.4, 4.5, 5.1, 5.2 and Chapter 6.

In situations where the uncontrolled variation can be considered to be an independent sample from a normal distribution there is a specified probability, usually chosen to be 95% or 99%, that the true value of the parameter is within $\pm\, t$ standard errors of the estimated value. This range within which the true value almost certainly lies is called the **confidence interval** of the parameter. The 95% confidence interval of the difference in the means for Example 3.2.1 is $1155 \pm (2.45 \times 554.8)$. The 2.45 comes from Student's t distribution. t is tabulated in many statistical textbooks and varies according to the number of degrees of freedom, or independent pieces of information, going towards the estimate of the variance of uncontrolled experimental variation. It is the number of measurements on independent experimental units minus the number of parameters estimated from the data. In Example 3.2.1 the degrees of freedom was six as two parameters have been estimated from the eight data points. The requirement for independence was emphasized in section 2.1 when the experimental unit was defined. The need for independence is now seen to be necessary in order to make unbiased estimates of the confidence interval for the parameter estimates.

One usual restriction on the uncontrolled variation is that it should be as if it were from a population with the same variance. This latter requirement has serious consequences in some common situations. Large responses are often more variable than small responses. In this situation, the use of least squares can result in serious under- or overestimates of the size of the parameter standard errors and there is no way of designing the problem away. It must be dealt with by analytical methods. Sometimes it can be overcome by transforming the data. Two common data transformations are to take the logarithm or the square root of the measurement. However, this often replaces one problem with another, as it is sometimes difficult to interpret the transformed data.

Repeated measurements on the same experimental unit – for example, a record of the growth of a group of animals – can be even more troublesome. With such measurements it is usual for the uncontrolled variation to be proportional to the size of the animal at the time the measurement is made. Measurements from time to time therefore do not have the same variance. In addition, it is unusual for measurements made on the same animals from time to time to be independent. Big animals have a tendency to stay big and small ones have a tendency to stay small. That is, the uncontrolled variation at any particular measurement time is dependent on the uncontrolled variation in the previous measurement.

One solution to this problem is to use a generalization of least squares, called weighted least squares. This technique allows for unequal spread of measurements and for the lack of independence between measurements. For more details of these techniques the reader is referred to Silvey (1975), Draper and Smith (1966) and Seber and Wild (1988). Other solutions are to describe the response by a summary statistic, a possibility already referred to in section 2.3. Another possibility involving the use of covariates is discussed in section 4.5.

There is a further important influence which affects the size of the standard errors of estimates of treatment parameters and predictions. It is the relationship between the chosen values of the predictor variates X at which observations are made and the standard errors. The Xs are sometimes called the design points. The design points are at 0, 30, 60, 90, 120, 150, 180 and 210 kg/ha in Figure 3.3.1. The design points have a large influence on the precision of the estimated parameters if the response is to be predicted from a continuous function such as $\hat{Y} = \hat{\alpha} + \hat{\beta}X$, $\hat{Y} = \hat{\alpha} + \hat{\beta}X + \hat{\gamma}Z$ or $\hat{Y} = \hat{\alpha} + \hat{\beta}\hat{\rho}X$. For example, the most precise parameter estimates of a linear function are obtained when design points are at the extremes of the range over which the function is to be used. Design points which cover the whole range of the exponential function provide precise estimates of all the parameters. If interest is in just one of the parameters other choices of design points are most efficient. For example, if interest is in obtaining the most efficient estimate of the asymptote, all the design points should give responses near the asymptote. Design points need to be carefully chosen in order to perform an experiment which is efficient for the purpose for which it is intended without using unnecessary experimentation.

Although the complexities of choosing suitable design points can be explored analytically, in many situations it is more easily achieved by computer simulation. A universal algorithm for such a simulation is as follows:

1. Select a real or true model which approximates the expected response. For example, $Y = 11\,000 - 2500 \times 0.980X$.
2. Generate values of Y for the chosen design points X.
3. By fitting the model to the data so obtained, it is possible on many programs to obtain the theoretical standard errors of parameters for any chosen **standard deviation** of random variation. If this is not possible proceed as follows.
4. Add a normally distributed random variate to the values of Y. Generating such random variation is an easily performed standard operation in many computer packages. The standard deviation of the random variation should be as would be expected for the uncontrolled variation from the experiment.
5. Fit the model by least squares. How to do this is described earlier in this subsection.
6. Repeat steps 4 and 5 one hundred times using different random samples from a normal distribution and take the average of the square of the reported standard errors for each parameter estimate.
7. Proceed through steps 2 to 6 for different selections of design points, varying both the number and their location. This could be for a predetermined grid of design points or for points chosen as a result of experience from preceding simulations.
8. Base the final selection of design points on the theoretical standard errors of the parameters if they are available; otherwise use the average of the square of the standard errors for each set of 100 simulations.

It is a wise precaution to ensure the final selection is robust against the range of possible true parameter values which are to be estimated. This is best done by varying the values of the true parameters in the first step of the above process.

Response surfaces which are used to model many industrial phenomena provide another example of the use of carefully chosen design points. In many industrial situations the response can be described by a curved surface similar to that displayed in Figure 3.3.2. In such a situation it is important to conduct a series of experiments so as to determine where the response is maximum and perhaps also where the response changes most rapidly. In the initial phase of an investigation it is most unlikely that an experiment will be conducted near the maximum and so a simple factorial experiment, each factor with two levels, is performed. A

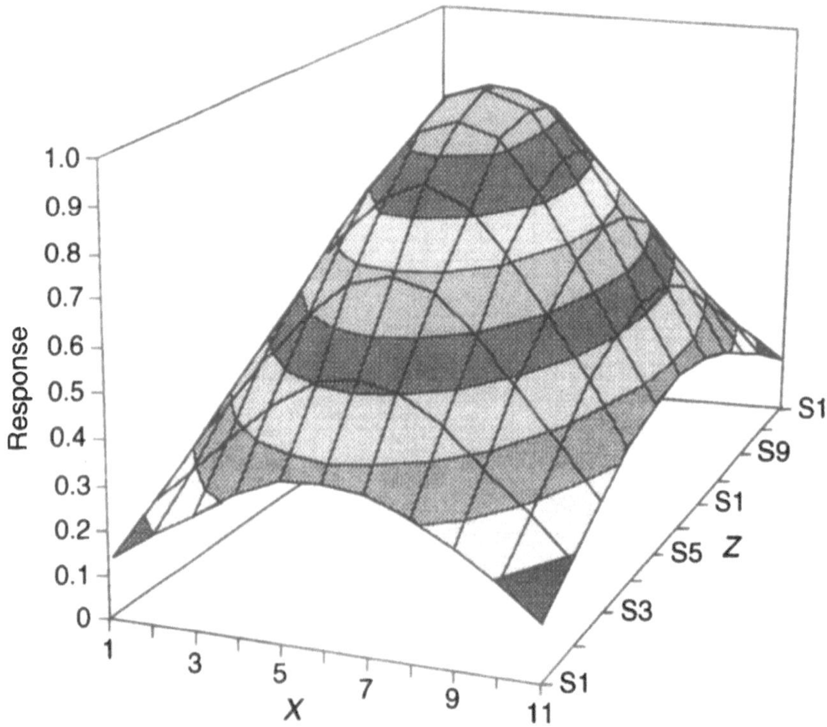

Figure 3.3.2 Response surface.

simple surface is then fitted to the data with the objective of finding the direction of steepest ascent. After a sequence of simple experiments the approximate location of the maximum is found and the data are then augmented by a further set of observations from carefully selected design points. This allows the modelling of the curved surface with low-order polynomials. Subsets of full factorial experiments and augmented factorial experiments are used in order to obtain parameter estimates and predictions of responses which have properties which are particularly desirable for the intended purpose. That is, the parameter estimates have high precision with a minimum number of design points. This allows the characterization of the surface with a minimum of disruption of a factor. Detailed accounts of the methods are given in Box and Draper (1987).

There is another situation where a carefully chosen set of design points leads to efficient experimentation. It was pointed out in section 3.2 that when the treatments are structured and there are a large number of factors, it is possible to model the response with interaction terms limited to those involving three or, in some situations, even fewer factors. Least-squares estimates of parameters for these reduced models can be made from a judicious choice of a fraction of a single replicate of the complete experiment. The penalty for adopting this strategy is that if interactions involving many factors are present, their effect is not distinguishable from interactions involving a smaller numbers of factors. That is, the interaction effects are said to be aliased. There is some notation which is often used in the description of fractional factorial experiments which is worth knowing. A 2^{k-p} experiment is a 2^{-p} replicate of a 2^k experiment. If it has resolution R, where R is a roman numeral, it means that interactions involving p factors can be estimated without being aliased by $R-p$ factor interactions. For example, a 2^{8-2} experiment is a 2^{-2} (= ¼) replicate of a 2^8 experiment. If it had resolution V, it would allow for any interactions involving two factors to be estimated without being aliased by interactions involving $R - p = 3$ factors. Books with details of the designs for these experiments are listed in Appendix A. GENSTAT, and DSIGNX (Biomathematics and Statistics Scotland, Edinburgh), generate such designs. The use of fractional factorial experiments is an attractive option in large factorial experiments as the experimental effort and management problem can be prodigious.

Estimates of least-squares parameters for models which are non-linear functions present additional difficulties. In some situations it is not possible to obtain the estimates. This is because of the size of the uncontrolled variation in the measurements. To illustrate this idea, consider five points covering the range of values given by the equation

$$Y = \alpha + \beta \rho X,$$

where $\alpha = 11\ 000$, $\beta = -2{,}500$, $\rho = 0.316$ and $X = 0, 0.5, 0.75, 1$ and 2. At $X = 2$ the yield is 10 750, only 250 short of the absolute maximum of 11 000. Random normal variation of varying size was added to the true values of Y, thereby simulating the addition of uncontrolled experimental variation. The parameters were then estimated from the simulated data by least squares. From 1000 such trials, the number of times an estimate of the parameter ρ was outside the acceptable range of (0,1) or a curve could not be fitted are shown in Table 3.3.1.

Therefore, before embarking on an experimental programme involving the estimation of non-linear parameters, the precision of data and the design points for which it is possible to obtain admissible parameter estimates need to be established. This can best be done by computer simulation where each individual situation is carefully examined.

3.3.2 MAXIMUM LIKELIHOOD

In contrast to least squares, which requires no knowledge of the supposed mathematical form of the distribution of the uncontrolled

Table 3.3.1 The dependence of unsuccessful fitting on the magnitude of the uncontrolled experimental variation

Standard deviation of added variation	Number of times outside range (0,1), or a curve could not be fitted
100	0
200	2
300	23
400	94
500	177
1000	420

variation, maximum likelihood requires such knowledge. The basic idea of maximum likelihood is to select model parameters which give the greatest probability to the observed events. More formally, the value of a function, known as the likelihood function, which is the distribution or frequency function of observations, is maximized by a suitable choice of parameters. The technique is very general with many applications, one of which includes parameter estimation for a class of models which are known as generalized linear models. These are generalizations of the linear models previously considered.

The fundamental idea of generalized linear models is to modify linear models so as to be able to deal satisfactorily with certain types of data which cannot be adequately dealt with by ordinary least squares. There are two reasons for this: first, the response can be described by a linear relationship only on a transformed scale; and second, the uncontrolled variation is dependent on the size of the observation and so cannot be considered to be a sample from a population with the same variance. Amongst the many types of data which are best dealt with by generalized linear models are **categorical data** and binomial data, both of which are discussed below.

As with least squares, there are many computer implementations, and generalized linear models are specifically catered for by GLIM (Numerical Algorithms Group, Oxford, UK). Implementations are also available in many other general specialist statistical programs such as GENSTAT, BMDP (BMDP Statistical Software, Los Angeles, USA) and SAS. Additionally, specific subsets of the class of generalized linear models are catered for by a multitude of specialist programs, usually with a limited scope and far too numerous to catalogue.

Categorical data

Categorical data consist of counts of measurements which fall into well-defined homogeneous groups or categories. An example is the categorizing of people according to their income group or hair colour, or the categorizing of an inflammatory response according to its severity. Each single category usually covers a range of possible measurements but for convenience the range is considered to represent a single class. For example, people whose income is less than NZ $15 000 might be classified under the single category of being very poor.

Predicted counts from the fitted model are given by

$$\hat{Y} = \exp(\hat{\tau}_i + \hat{X}_j)$$

where $\hat{\tau}_i$ is the estimate of the parameter describing the effect on the counts of the ith treatment and $\hat{X}_j$ is the estimate of the effect on the counts of the jth classification. The treatments might be the effect of various preparations on disease lesions which are categorized according to well-defined rules. The model-fitting process also obtains standard errors for $\hat{\tau}_i$ and $\hat{X}_j$ and from them confidence intervals for both $\hat{\tau}_i$ and $\hat{X}_j$ and the $\hat{Y}$.

Binomial data

Binomial responses consist of proportions of measurements which respond to an applied stimulus. Such responses to increasing stimuli are invariably best described by a sigmoid-shaped curve. An example is the proportion of insects dying with increasing doses of insecticide. The predicted number out of n insects responding is given by the fitted model as

$$\hat{Y} = n[\exp(\hat{\eta}) / (1 + \exp(\hat{\eta}))],$$

where the predicted probability of responding is given by the expression involving $\hat{\eta}$ which itself is a linear model such as $\hat{\eta} = \hat{\tau}_i$, $\hat{\eta} = \hat{\alpha} + \hat{\beta}X$ or $\hat{\eta} = \alpha + \hat{\beta}X + \hat{\gamma}Z$.

As with linear models where least squares is a satisfactory strategy for estimating the parameters, the choice of design points with generalized linear models strongly influences the magnitude of the standard errors of the parameter estimates. Consequently, a similar process to that used for least squares for designing such experiments is useful.

A simple algorithm for a design for a binomial response is as follows:

1. Select a real or true model which approximates the expected response, but on the transformed scale. For example, $Y = n[\exp(\eta)/(1 + \eta)]$, where $\eta = -61 + 35X$ for $X = 1.5$ to 1.9.
2. Generate values for Y, the number responding out of n for each of the chosen design points X, and calculate the binomial proportion Y/n.
3. By fitting the model to the data so obtained, it is possible on many programs to obtain the theoretical standard errors of parameters. If this not possible proceed as follows.

4. For each Y/n, generate a random sample of n from a binomial distribution with mean Y/n. Generating such random samples is an easily performed standard operation in many computer packages.
5. Fit the model.
6. Repeat steps 4 and 5 one hundred times using different random samples from the binomial distributions each with mean Y/n and take the average of the square of the reported standard errors for each parameter estimate.
7. Proceed through steps 2 to 6 for different selections of design points, varying both the number and their location. This could be for a predetermined grid of design points or for points chosen as a result of experience from preceeding simulations.
8. Base the final selection of design points on the theoretical standard errors of the parameters if they are available; otherwise use the square of the average standard errors for each set of 100 simulations.

As with least squares, it is a wise precaution to ensure the final selection is robust against the range of possible true parameter values which are to be estimated.

A more complete account of the application of generalized linear models for categorical data and binomial data is contained in Dobson (1990), and the extensive additional repertoire available with generalized linear models is discussed in McCullagh and Nelder (1989). General accounts of the method of maximum likelihood are contained in many statistics textbooks, including Silvey (1975) and more recently Stuart and Ord (1991).

3.3.3 RESIDUAL MAXIMUM LIKELIHOOD

Residual maximum likelihood is sometimes called restricted maximum likelihood or REML. REML will fit the same models as least squares. However, it will also fit models in situations not possible with least squares. It will fit models which have more than one identifiable source of uncontrolled variation and estimate the size of each source without the severe limitations imposed on the experimental layout by least squares. The estimates of the three sources of uncontrolled variation, known as the components of variance, in Example 2.1.2 were made using REML. The data used for these estimates came from commercial flocks which were not all the same size. Another example where there are three sources of variation is from plant variety testing, where there is variation from year to year, between experimental sites and between experimental units within sites and years. REML is used extensively when the strategy of blocking is used in the laying out of an experiment. Blocks are sets of experimental units which are homogeneous in some sense; their use is discussed in section 5.2. They make it possible to improve experimental precision by modelling the heterogeneity between the homogeneous blocks. In classical experimental design theory it was necessary for the blocks to be all the same size. This often unrealistic restriction is lifted by the use of REML as it allows the modelling of the effect of blocks of natural but disparate size.

Computer implementations are available in both GENSTAT and SAS. As with least squares, the use of templates set up by experts and modified by novices with minimal instruction is advocated for non-expert users.

Details of the REML technique are given in Patterson and Thompson (1971) and Thompson (1977). A more accessible account is given by Robinson (1987).

3.4 ASSESSING THE ADEQUACY OF MODELS

The ultimate and only real test of the adequacy of a model is its ability to give good predictions. 'Good' is a subjective description and must be judged in the light of the stated experimental intentions. Good for one application may be hopelessly inadequate for another. For example, the standard errors of the mean responses obtained from a simple additive effects model might be too large to enable them to be used as a useful predictor of the financial outcome of the application of an alternative treatment while still allowing a confident conclusion that alternative treatment resulted in a substantially different response. Care is needed to distinguish between good predictions and good-looking predictions. Flashy computer outputs can deceive a user into thinking that a poor predictive model is in fact very good. Of course the reverse is also true.

Some examples of inadequate models include the following:

1. The model does not adequately describe the situation which is known to exist: this results in frankly incorrect predictions; its presence is frequently disguised by elegant and sophisticated computer displays.
2. There is a large component of variation in the modelled phenomenon which is not accounted for: Example 2.1.2 concerns a situation when variation between groups of animals at different sites is larger than variation between individuals within a site. Therefore, if between-site variation is not allowed for when considering the standard error of predictions, the predictions may be misleadingly inaccurate for any particular site.
3. A model is used in a situation for which it is not intended: it would be unreasonable to expect a model for the dynamics of a fish population from one part of the world to be adequate for another species in another part of the world without some adjustment.

4. A model is extrapolated beyond the range for which it has been fitted: if a quadratic equation is used to describe an upward response, use of the equation where predictions plunge rapidly through to negative values would be seriously misleading.

Care is needed in assessing models for each of the above deficiencies.

When models have been fitted by statistical techniques, a disciplined methodology is available to help assess their adequacy. It is in two parts.

3.4.1 DOES THE FUNCTION DESCRIBE THE DATA ADEQUATELY?

A simple and effective technique used to assess the adequacy of a fitted model is to look at a graph of the model predictions, on which are also plotted the measurements used to estimate the parameters of the model. Discrepancies between the measurements and the predictions are readily apparent to the eye. Visual perception is good at detecting even subtle discrepancies. This is demonstrated in Figure 3.4.1, where random uncontrolled variation with a standard deviation of 500 has been added to the true values of the equation $Y = 11\,000 - 2500 \times 0.980X$ for values of X in steps of 30 between 0 and 210, and the data so generated used to obtain estimates of the parameters of the generating equation. As a demonstration of a visually apparent inadequate model the same data have been used to estimate the equations of a straight line, that is $\hat{Y} = \hat{\alpha} + \hat{\beta}X$. The predicted or estimated response from each of the two fitted equations, firstly the adequate fit and then the inadequate fit, are displayed as continuous lines in the two graphs of Figure 3.4.1. The diamonds represent the generated data.

This technique can be complemented by looking at a graph of the differences between the data and the predictions from the fitted model. These so called residuals, when plotted against the predictions, provide a visual means of detecting discrepancies between the data and the fitted model. When least squares has been used for the fitting process and the model being fitted is adequate, the residuals appear in a band of uniform width for all predicted values. Systematic deviations from this pattern indicate an inadequate model. For example, one would expect more residuals with negative signs for small and large predicted values, and more residuals with positive signs for mid-range predicted values, in the second of the two fitted curves in Figure 3.4.1. This is in fact what happens as is displayed in Figure 3.4.2.

Inspection of graphs like Figure 3.4.2 can also reveal another aspect of the fit which is worthy of consideration. The residuals might indicate that the model does not adequately describe the data. This would be the case if some of the measured points were too far away from the predictions to be of value. This idea is formalized in the statistic called the residual standard deviation, which gives a measure of the spread of the

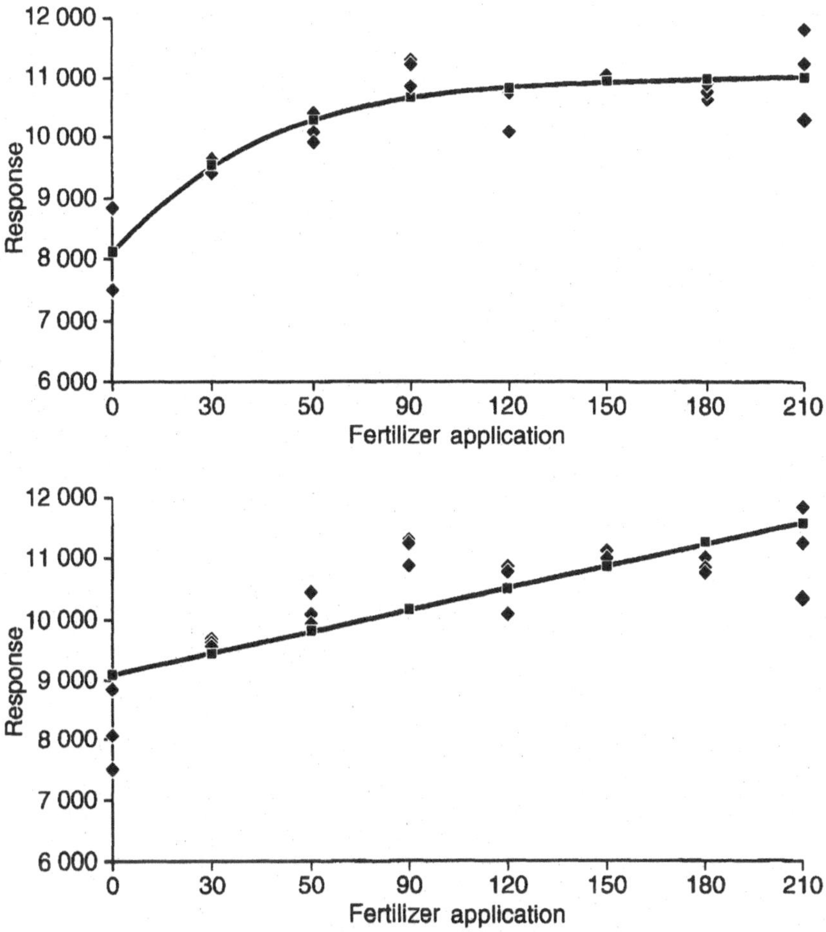

Figure 3.4.1 A visually apparent adequate fit (top) and a visually apparent inadequate fit (bottom).

residuals. It is calculated from the square of the differences between the measurements and the predicted values.

Figure 3.4.3 displays what the plot of residuals on predicted values might look like if the uncontrolled variation increases as the magnitude of the measurements increases. The characteristic fan shape of the residuals as the model predictions increase is indicative that an alternative analysis involving data transformation or the application of generalized least squares should be considered.

If the data came from a spatial arrangement such as experimental field plots, it is good practice to plot the residuals, or simply the sign of the

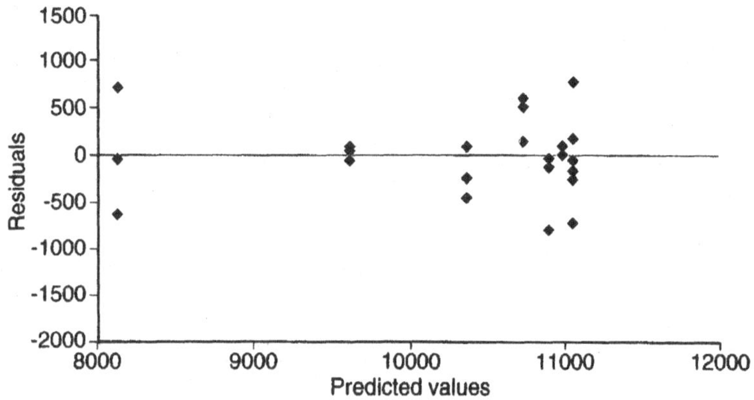

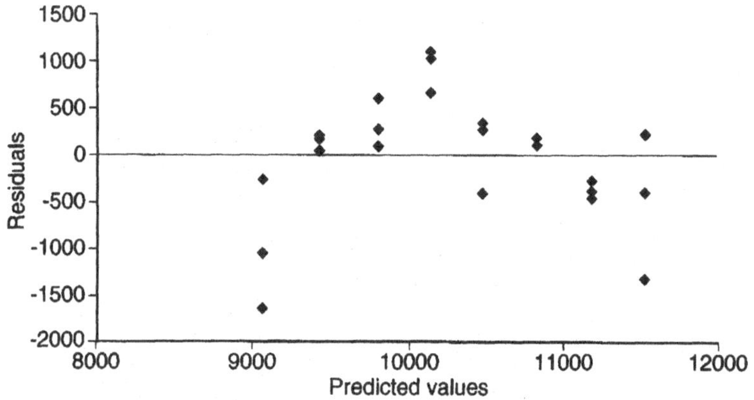

Figure 3.4.2 Residuals plotted against fitted values for an adequate fit (top) and a similar graph for an inadequate fit (bottom).

residuals, on the field plan. Environmental effects which were not visible when a field experiment was planned, laid down or harvested can be found by this simple technique. The suspected environmental effects can then be modelled along with the model of the treatment effects. Such techniques are described in sections 4.5 and 5.2. An environmental gradient has been generated by simulation and is shown in Table 3.4.1. Each value of the series of uncontrolled variation is dependent on its predecessor. That is, positive values tend to be followed by positive values and negative values tend to be followed by negative values. Consequently the signs of the residuals do not change as frequently as would be expected if the dependence did not exist.

Similarly, if the data are a time sequence, plotting the residuals in the same order as the data were collected reveals dependencies between the

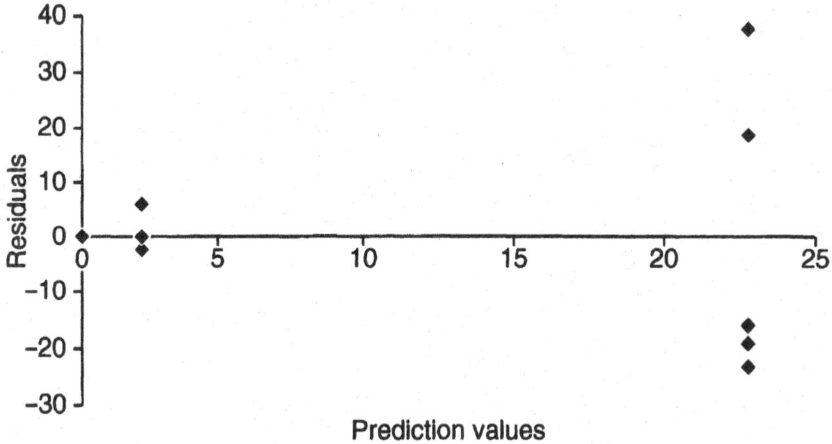

Figure 3.4.3 Residuals plotted against predicted values when uncontrolled variation increases in magnitude as the size of predictions increase.

Table 3.4.1 A plot of the sign of residuals on a field plan when the observations are not independent

+	+	−	−	−	−	−	−	−	−	−	+	+	+	+	+	−	−	+	+	+

residuals if such dependencies exist. Frequently, the difficulties associated with such dependencies are avoided by calculating variates which are relevant summaries of such sequences, a process which was described in section 2.3. An example of a time sequence of measurements where each measurement is dependent on previous measurements is shown in Figure 3.4.4.

The adequacy of a model which is a linear or non-linear function can be checked by having at least one more design point than the number of parameters which are to be estimated from the data. It is then possible to check if the variation in the measured response not accounted for by the model is bigger than would have arisen from uncontrolled variation in the measurements. This would be an indication that the fitted model did not adequately describe the measurements. For example, if it was required to fit the linear model $\hat{Y} = \hat{\alpha} + \hat{\beta}X$ or the non-linear model $\hat{Y} = \hat{\alpha} + \hat{\beta}\hat{\rho}X$ at least three design points for the first and four design points for the second would allow a check on the adequacy of the fitted model. An example is shown in Figure 3.4.5.

Suppose that the three points are the means of the responses from replicated measurements on the predictor variable. Also suppose that the

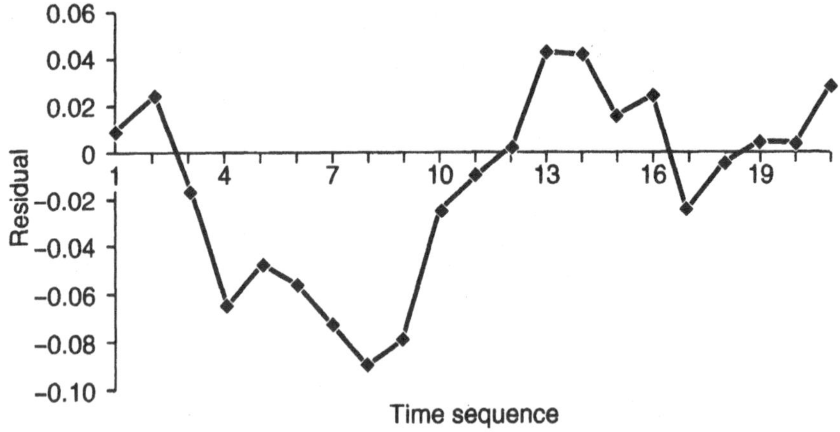

Figure 3.4.4 Plot of dependent residuals from a time sequence of measurements.

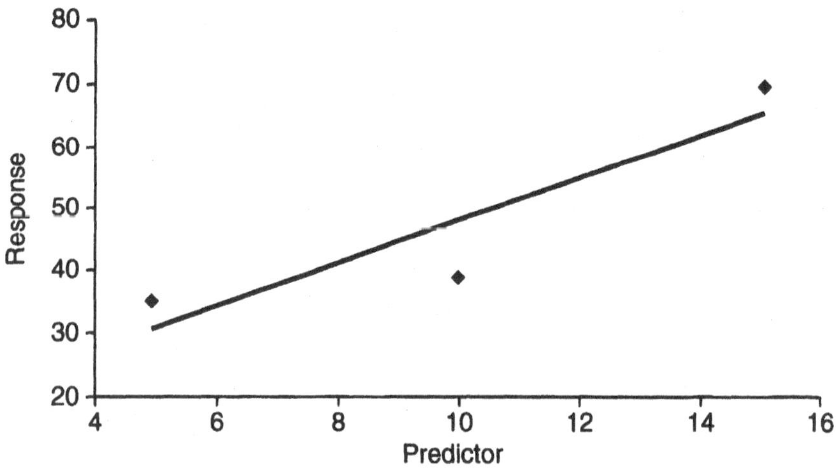

Figure 3.4.5 A visually apparent inadequate model.

deviation of the mid-point from the line joining the two end-points is larger than would be accounted for by the estimate of uncontrolled variation in the data points. Then the inadequacy of the straight line as a predictor of the response is clearly demonstrated. If, before the experiment from which the above data were obtained, it had been supposed that a straight line was an adequate model for the response and only the

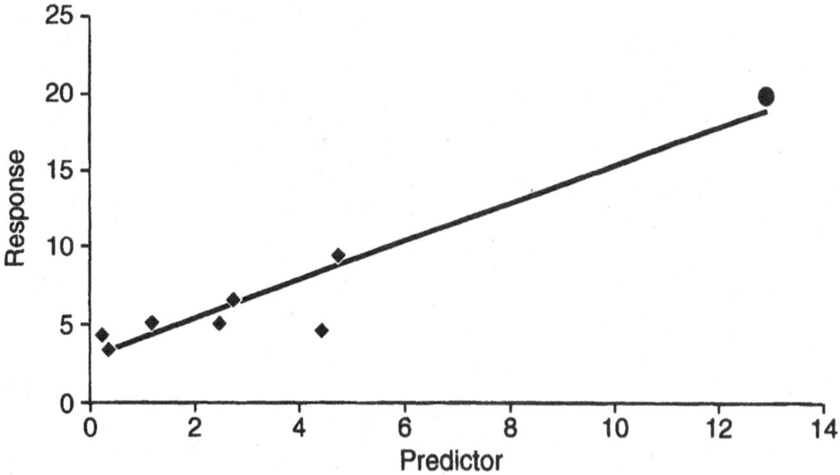

Figure 3.4.6 A point of high leverage (dot).

two end-points were measured, the evidence from the mid-point concerning the inadequacy of the straight-line model would have been missed.

There is a bewildering array of test statistics which are useful additions to graphs of the data and residuals on the fitted values. Statistical packages report many of them automatically.

A point with high leverage is a point which has an unusual predictor. A high leverage point exerts a large influence on the size of its predicted value. The point identified by a dot as opposed to a diamond on Figure 3.4.6 is such a point. If the measured response were moved down towards the X axis the slope of the line, and its predicted value, would move down with it. The term 'high leverage point' for such a point is well chosen. It operates in the same way as a lever with a fulcrum at the mean of the predictor (X) variate.

If when model parameters are re-estimated with points of high leverage excluded from a data set the parameter estimates change markedly, a modeller would have to be very confident of the integrity of those data points before placing much faith in the estimated parameters. Consequently it is important to design experiments so as not to depend on just one design point for the estimation of an important parameter. In Figure 3.4.6 it would be much better to have an equal number of design points at each end and in the middle of the range of the predictor variate, thereby reducing dependence on just one datum point as well as providing a check on the adequacy of the straight line to model the relationship.

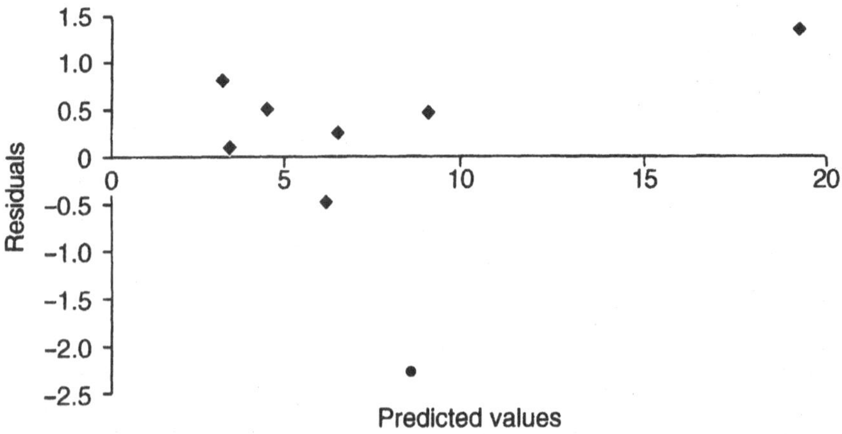

Figure 3.4.7 A point with a large residual (dot) from the model depicted in Figure 3.4.6.

A point with a large residual is one which is a long distance from the predicted or fitted value for that point. A high residual identifies an unusual response. When modelling, points with large residuals should be carefully reviewed. Large residuals can indicate an error in the recording or entering of a datum. They can also indicate inadequacies in the chosen model. The point indicated by a dot in Figure 3.4.7 has a high residual value where the residuals are from the model and data presented in Figure 3.4.6.

There are a number of ways in which residuals are reported. They can be reported as the discrepancy between the predicted value and datum point as in Figure 3.4.7, or as either standardized or Studentized residuals. Standardized or Studentized residuals are scaled so as to make values outside the range (–2, 2) worthy of investigation.

Cook's distance combines the information about leverage and standardized residuals of each point, and a high value suggests an experimenter consider that particular point carefully.

R^2 values are further diagnostic statistics. There are two statistics which are referred to as R^2 values. Both vary between 0 and 1, with a value near 1 being seen as evidence of a good fit of the model to the data. Both are sometimes expressed as a percentage, and both indicate the proportion or percentage of variation in a set of data which is accounted for by a fitted model. One of the R^2 values is sometimes called the coefficient of determination but more often it is known simply as R^2. It is calculated as $\Sigma(\hat{Y} - \bar{Y})^2 / \Sigma(Y - \bar{Y})^2$, where Y are the measurements, $\hat{Y}$ are the corresponding predictions from the fitted model and $\bar{Y}$ is the mean of

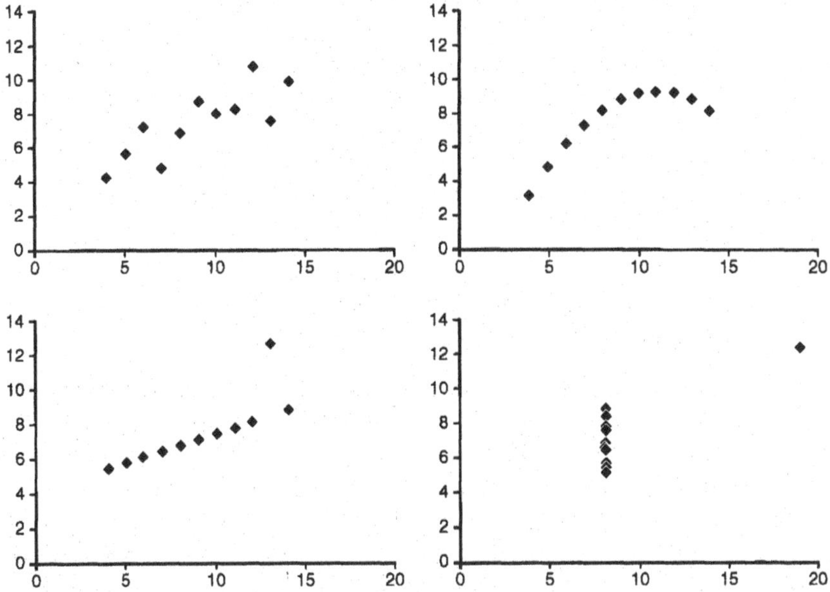

Figure 3.4.8 Four data sets to which a linear equation was fitted and where the estimated parameters and R^2 values are all the same.

the *Y*s. Its value from the model presented in Figure 3.4.6 is 0.91. The other R^2 is called adjusted R^2. It is $1 - S_e^2/S_Y^2$ where S_e^2 is the estimate of the variance of the uncontrolled variation after the model has been fitted and S_Y^2 is the estimate of variance of the measurements before any model fitting has taken place. Its value from the model of Figure 3.4.6 is 0.898. Either one or the other, and it is often not clear which one, is frequently quoted in journal papers. It really does not matter much as an R^2 value is sensibly interpreted only in conjunction with the other diagnostic methods so far discussed. An effective demonstration of this assertion was made by Anscombe (1973). Figure 3.4.8 displays four data sets to which a linear equation, $\hat{Y} = \hat{\alpha} + \hat{\beta} X$, was fitted. The estimated parameters of the fitted line for all four were the same; and so are the R^2 values, all of which are equal to 0.67.

So called 'tests of good fit' are often used in the model-fitting process. They require cautious interpretation. This is because the magnitude of the test statistic, the value which is compared to the tabulated values in statistical tables, can depend on arbitrary choices. An example is the Pearson χ^2 test based on calculating $\Sigma(O-E)^2/E$, where O and E stand for observed or measured and expected or predicted frequency of measurements in arbitrarily chosen categories or classes. It is possible to manipulate the value of the statistic by judicious choices of the class boundaries and hence produce a result which the manipulator finds desirable. In addition, the interpretation of a test of good fit requires careful thought. A non-significant result which can be taken to imply a good fit does not necessarily mean the fit is good for the purpose the modeller has in mind. Conversely, a significant result which can be taken to imply a bad fit does not necessarily mean the model is not adequate for the purpose for which it is intended.

3.4.2 WHICH IS THE BEST CHOICE FROM A COMPETING SET OF POSSIBLE MODELS?

Choosing between models is an important and difficult task. Because it is difficult, it is tempting to look for a single statistical test which it is hoped is objective and which takes away the modeller's responsibility for making the choice. Although many such tests have been proposed, none is able to achieve the aim. The reason is that the choice between alternatives is conditional on what the model is required for. Choosing a model on the basis of changes in or of the magnitude of the R^2 values is not necessarily a method which selects the best model. The differences in R^2 values which are used to discriminate between models may be due to uncontrolled variation and so may lead to unimportant improvements in predictions. Choosing the model with the largest R^2 value can also lead to large standard errors of predictions because of the inclusion in the model of imprecisely determined parameters. It may also lead to the choice of models which are not as theoretical or intuitively appealing as some of the competing alternatives. Graphical procedures, coupled with the idea of building on accepted theory concerning the system being investigated, are usually the most satisfactory selection criteria. There is no single test criterion which can be universally applied when choosing between models. It is a difficult task where decisions have to be made by the modeller.

How to make comparisons

4

4.1 IDENTIFYING COMPARISONS WHICH NEED TO BE MADE

In order to interpret experimental measurements, and choose between possible alternative models, it is necessary to make comparisons. These comparisons are usually between the estimated or predicted responses of just two treatments. Frequently, however, these two estimates are from a much larger experiment which provides an opportunity to make more elaborate and general comparisons, sometimes known as contrasts. The following example illustrates the idea.

EXAMPLE 4.1.1

An experiment was undertaken in order to investigate how much the use of a selenium supplement in sheep increased the production of wool. Four treatments were to be compared. One was a control in which no selenium was administered. The other three treatments were various ways of administering the selenium to sheep – one of these dose forms provided instantaneously available selenium and the other two were slow-release dose forms. The investigator formulated three contrasts to address the intentions of the experiment.

First, what was the response to selenium? If selenium was effective in increasing wool production then the animals given the three selenium dose forms would all be expected to yield more wool than the untreated control animals. It is reasonable, therefore, to add together the three separate comparisons between the three selenium dose forms and the control to give an omnibus type of comparison. This is what has happened to produce the first contrast in Table 4.1.1, in which the sum of the yield from the three dose forms is compared with three times the yield from the controls. This contrast would be expected to provide an estimate of the overall response to selenium.

Table 4.1.1 Contrasts between a set of treatments

Contrast	Selenium treatment			
	Control	Instant	Slow 1	Slow 2
1	-3	1	1	1
2	0	-2	1	1
3	0	0	1	-1

Second, what was the difference between the response to the instantaneously available dose and the slow-release dose forms? If the slow-release form was more effective than the instantaneous dose form then two slow-release dose forms would be expected to yield more wool than the instantaneous dose form. As before, it is reasonable to add together the two comparisons between the two slow-release dose forms and the instantaneous dose form. This leads naturally to a comparison between the instantaneous available dose form and the two slow-release dose forms as shown in the second contrast in Table 4.1.1.

Third, what was the difference in response between the two slow-release dose forms? This leads to the usual straightforward comparison between the two slow-release dose forms.

The strategy of adding together similar simple comparisons to make more complicated contrasts increases the probability of detecting differences between treatment effects by increasing the replication of a comparison. The reason for this satisfactory outcome is discussed in section 4.2.

Contrasts are used in many situations, and it is useful to use them to describe estimated responses of factorial experiments. Box and Draper (1987) and Mead (1988) provide details of some of the many applications. An example of the way contrasts are used in factorial experiments is displayed in Table 4.1.2. There, contrasts from an experiment involving two levels of nitrogenous fertilizer, n_0 and n_1, two levels of phosphate fertilizer, p_0 and p_1, and two levels of potassium fertilizer, k_0 and k_1, are displayed.

The table is interpreted as follows. The contrast N is simply a comparison between the responses in plots with levels n_0 and n_1 of nitrogen. Notice that the contrast between n_0 and n_1 is made within each of the four treatment combinations, p_0k_0, p_0k_1, p_1k_0 and p_1k_1. The NP contrast is obtained as the product of the contrasts defined by N and P. It may be arranged into the following expression which facilitates interpretation:

$$[n_0(p_0-p_1) - n_1(p_0-p_1)](k_0+k_1).$$

The first term of this contrast, $n_0(p_0-p_1) - n_1(p_0-p_1)$, is the difference in levels of phosphate (p_0-p_1) in the presence of n_0 minus the difference in

Table 4.1.2 Contrasts between a set of treatments from a factorial experiment

Contrast	Treatment combination							
	$n_0p_0k_0$	$n_1p_0k_0$	$n_0p_1k_0$	$n_1p_1k_0$	$n_0p_0k_1$	$n_1p_0k_1$	$n_0p_1k_1$	$n_1p_1k_1$
$N = n_0 - n_1$	1	−1	1	−1	1	−1	1	−1
$P = p_0 - p_1$	1	1	−1	−1	1	1	−1	−1
$K = k_0 - k_1$	1	1	1	1	−1	−1	−1	−1
NP	1	−1	−1	1	1	−1	−1	1
NK	1	−1	1	−1	−1	1	−1	1
PK	1	1	−1	−1	−1	−1	1	1
NPK	1	−1	−1	1	−1	1	1	−1

levels of phosphate in the presence of n_1. That comparison is made for each of the other two experimental treatments, k_0 and k_1. Consequently the difference between the estimated responses in the plots indicated by the expression is interpreted as an estimate of the interaction between nitrogen and phosphate. Similar interpretations of the other contrasts notationally represented by *NK*, *PK* and *NPK* can be made. The *NPK* contrast is an estimate of how the interaction between any two of the factors differs at each of the two levels of the third factor.

In any experiment involving unstructured treatments there are two different types of information an experimenter might want to know. In everyday language these are often expressed as follows. Are contrasts amongst the treatments different, or are they, for the purposes at hand or for practical purposes, the same? These questions can only be answered after careful definition of the meaning of 'different' and 'the same'.

For most situations, the real meaning of treatments being 'different' is as follows. The experimenter requires to establish that a difference in responses to different treatments is as big as, or bigger than, a specified quantity. This quantity is often not verbalized or written down, but should be present in the experimenter's mind. In a well-designed experiment it is necessary to specify this quantity or bound. Expressed more succinctly, the experimenter is required to establish a bound for the size of the difference between the responses to particular treatments. The bound is a quantity chosen above which it is required to detect the difference despite the masking effect of uncontrolled variation. An example would be to establish a bound above which the difference in responses to different formulations of the same drug almost certainly lay. This would be done when there was reason to believe an important clinical difference would result if the difference in responses was as big as, or bigger, than specified. In a farming situation a suitable bound for the size of the response to a management strategy is the response which would more than cover the cost of the application of the strategy.

In a similar way the interpretation of treatments producing 'the same' effect is expressed as follows. The difference in the size of the response to particular treatments is smaller than a specified bound. The specified bound is a quantity chosen below which it is not necessary to detect the difference over the masking effect of uncontrolled variation. An example would be to specify a difference in response to two drug formulations below which they would be considered to be bio-equivalent. That is, the outcome for a person or animal taking the drug would be considered to be not materially affected by the choice of formulation. In a farming situation a bound for the size of the response to a management strategy is the response which would not cover the cost of the application of the strategy.

It is emphasized that both these types of requirements are conditioned by the words 'specified bound'. Such conditioning is rarely stated but should be an essential part of the designing of all experiments. The specified bound is the quantity above which it is important to detect any difference in response or below which it is not important to detect any difference in response. If it is not specified the point of designing the experiment can be lost because the experiment may have a very small probability of detecting important differences which a failure to detect may be falsely misinterpreted as there being no difference. This is one of the most important failings in the interpretation of experiments. Alternatively, the experiment may have a high probability of detecting differences much smaller than are of interest. This results in waste of experimental resources and, if animals are involved, potentially extravagant ethical costs.

When the treatments are structured it is often required to model the response by a continuous function. This could be the slope of a straight line or a parameter or set of parameters describing some aspect of a more complicated function. An example is the by now familiar exponential curve describing the yield response from a sequence of rates of fertilizer application.

In this situation the two types of information an experimenter might want to know are analogous to the information an experimenter might want to obtain from unstructured treatments. It might be required to establish a bound for the size of the difference between particular parameters of functions which describe responses to different treatments. For example, two fertilizers might be considered to be different if they produce responses with asymptotes which are greater than, say, 200 kg/ha apart in the first year of application. In this case 200 kg/ha is a suitable bound.

Similarly, an experimenter might want to conclude that the size of the difference between particular parameters of functions which describe responses to different treatments is below a specified bound. For example, two fertilizers might be considered to be the same if they produce responses with asymptotes which are within, say, 100 kg/ha of each other in the first year of application. In this case 100 kg/ha is a suitable bound.

As with unstructured treatments, care is needed in designing experiments which are capable of detecting the specified differences. Failure to take the necessary care in designing such experiments can lead to important differences being undetected or an extravagant use of resources. Techniques to facilitate the designing of powerful, but not extravagant, experiments for specified differences in responses are dealt with in the next section.

4.2 POWER OF COMPARISONS

Comparisons which are intended to find specified differences need to be made in such a way as to avoid the masking effect of the uncontrolled variation. That is, comparisons need to be powerful. Comparisons which lack power can have quite substantial differences obscured by uncontrolled variation. The natural way to understand the concept of power is to understand statistical **significance tests**. The logistics of how to perform a significance test can be found in most standard statistical texts. However, what is the interpretation of a significance test? The computer simulation described below helps answer this question.

1. The simulation generates two normally distributed populations each of many individuals. These can be thought of as the results of all possible outcomes from a two-treatment experiment. The population means correspond to the real or true responses τ_i and the deviations of individual observations from the mean correspond to the ε in the model, $Y = \tau_i + \varepsilon$, which was discussed in section 3.2. The deviations from the real or true responses, the ε, have the same dispersion of values or variance, which is arbitrarily chosen to be 1. For each simulation it is possible to make the population means different but initially they are both set equal to 0.
2. At each simulation, a random sample of ten so-called replicates is drawn from each population and the sample mean, the $\hat{\tau}_i$, and the standard error of the difference between the two sample means is calculated.
3. From these a 95% confidence interval for the difference in the population means based on the measured difference between the two sample means and the standard error is obtained. If the 95% confidence interval includes 0 the sample means are not significantly different. Otherwise they are said to be significantly different at the 5% level. This interpretation of a significance test is easily shown to be algebraically equivalent to other perhaps more familiar formulations.

4. Steps 2 and 3 are repeated one hundred times and the number of statistically significant results at the 5% level is counted.

This process allows an assessment of what happens in situations when the population means differ by various amounts. The first interesting situation is when the population means are both the same. There is a probability of 95% that the population mean is within the range specified by the confidence interval. Therefore, when the population means are both the same it would be expected that in the long run the confidence interval would not include zero 5% of the time. The simulation shows that is in fact what happens. When the confidence interval does not include zero an error of the first kind, or a type 1 error, has been made. That is, it will be in fact concluded there is a significant difference between the two population means where there is no difference. There is no way of overcoming this difficulty.

The next interesting situation is when the population means differ. The simulation shows that, as the size of the difference between the population means increases, the number of significant results increases. However, even when real population differences exist, non-significant results still occur. Importantly, therefore, it is not possible to conclude when there is a non-significant result that real differences between populations do not exist. When there is a real difference between the population means, and the test is not significant, an error of the second kind, or a type 2 error, has occurred. This occurs often. It is hard to envisage many situations with comparative experiments when a real or true difference between the responses to different treatments can be expected to be exactly zero.

Note that a type 2 error does not necessarily mean the experiment is inadequate in some way. It may simply mean that the experiment was not designed to detect such small differences, presumably because small differences were of no interest. However, experiments should be designed to have a high probability of obtaining a significant result when a predetermined and specified real difference in the response to treatments is present. In other words, the power of the experiment, for the purpose for which it is designed, should be high. The predetermined and specified real difference is the specified bound of the previous section.

In the above example the probability of obtaining a significant result at the 5% level from a known difference in the population means, that is, the power of the experiment, is plotted in Figure 4.2.1. The number of significant results can be increased when a real or true difference is present simply by increasing the number of replicates in the experiment. This increases the power of the experiment. An example where the true differences between the population means is the same as one standard deviation of either population is displayed in Figure 4.2.2.

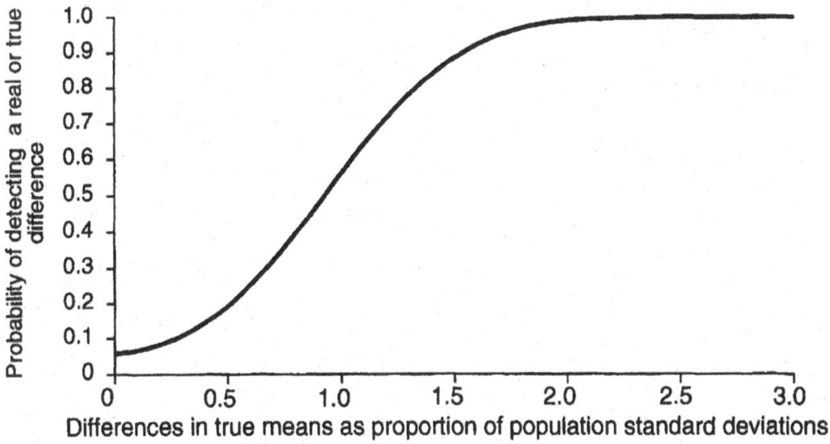

Figure 4.2.1 Power curve for the computer simulation experiment.

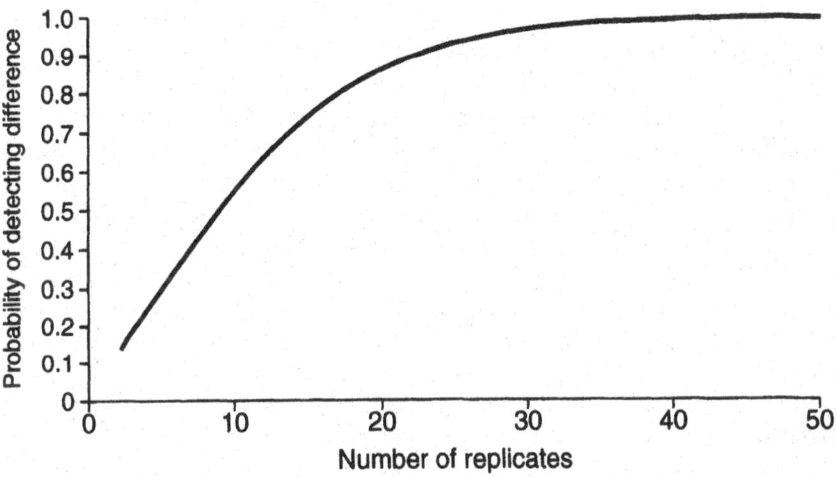

Figure 4.2.2 Power curve for the true difference equal to the population standard deviation.

In Figures 4.2.1 and 4.2.2, attention was focused on determining the power of a comparison for a specified true difference and number of replicated measurements. It is possible to work in another way. That is, to determine the number of replicates required to detect a specified true difference and probability of detecting that difference. When used this way, power becomes a formidable experimental design tool. Using a 5%

Table 4.2.1 The dependence of the power of comparisons on the relative size of the response

Difference as proportion of SD	Number of replicates for a 90% chance of detecting the difference
0.25	340
0.50	86
0.75	39
1.00	22
1.25	15
1.50	10
2.00	7
3.00	4

level of significance, Table 4.2.1 shows the number of replicates required to ensure an experiment has a 90% chance of detecting the specified differences, which are expressed as a proportion of the standard deviations (SD) of individual measurements. It will be noted that the smaller the true difference the more replicates will be required to obtain a given proportion of significant results. Other probabilities frequently used for detecting differences are 80%, 95%, 99% and 99.9%.

In order to exploit the idea of power as a design tool it is not necessary to simulate each situation. The use of Table 10 from Pearson and Hartley (1970) facilitates the calculations. Alternatively, there are computer programs available which are easy to use and provide rapid answers. An example is the GENSTAT procedure POWER presented in Appendix C.

Power calculations require the real or true variance (or its square root, the standard deviation) of the population, a sample from which is to provide the experimental units. This real or true value is never known except in hypothetical situations such as that described in the above computer simulation. Therefore, in order to use the idea of power as an experimental design tool, it is necessary to make plausible guesses of the population's real or true standard deviation. The idea of making plausible guesses is sometimes balked at but in reality presents few difficulties. If no information about the standard deviation is available the power calculations can be done for a range of possible alternatives. The results can then be used as a basis for an informed decision on the number of experimental units required. However, experimenters rarely have no information on the size of the standard deviation of their measurements as they have experience of previous experiments which have provided such estimates. Additionally, estimates of the standard deviation can be obtained from a knowledge of the expected range of measurements. Experimenters often have good intuition about such ranges. If the expected ranges of measurements are available, Table 22 from Pearson and Hartley (1970) can then be used to convert the ranges easily into estimates of standard deviations.

In the above example, comparisons between sample treatment means have been considered. However, exactly the same logic applies to more complicated contrasts and parameter estimates calculated from the data. Examples of such parameter estimates include the estimate of the slope of a straight line, a parameter of an exponential curve or an area under a curve. Power for these applications can sometimes be calculated by analytical methods. However, it is now more usual for them to be done by computer simulations where a random sample from a known distribution is added to a known treatment effect which is then estimated from the simulated data in an exactly analogous way to the simulation described above. The following example illustrates the idea.

Example 4.2.1

An animal is allocated to receive one of two treatments and after a specific time the response is classified into one of four categories. Data from such an experiment would conveniently be entered into the cells of the following table:

	Classification			
	1	2	3	4
Treatment 1				
Treatment 2				

Suppose the experimenter is required to establish that the treatments produce a different pattern of response. In order to proceed with a design, the experimenter would have to decide on what difference in pattern of response in the population (not the sample of animals used in the experiment) it would be important to detect. That is, the aim is to establish the bounds as discussed in section 4.1.

Suppose treatment 1 was a standard and there were some long-run figures available which would suggest an expected frequency of response resulting from its application of 30%, 40%, 20% and 10% of measurements classified as 1, 2, 3 and 4, respectively. Also suppose that, on the basis of some theory, treatment 2 is to be applied and is expected to result in a frequency of measurements in classes 1, 2, 3 and 4 of 0%, 50%, 30% and 20%, respectively. If the responses to the two treatments are different in the theorized way, a treatment by classification interaction will exist. It is therefore required to conduct an experiment which is capable of detecting this interaction. One method of detecting the interaction is as follows.

After the fitting of a generalized linear model to the categorical data as discussed in section 3.3, the differences between the fitted values and the observations, that is the residuals, are used to calculate a quantity known as the residual deviance. In this example involving just two treatments and four classifications, the residual deviance is a χ_3^2 random variable provided the model fits the data adequately. From a set of statistical tables it is seen that a χ_3^2 random variable will exceed a value of 7.81 only 5% of the time. Therefore a residual deviance greater than 7.81 suggests that the residuals are larger than would be accounted for by just random variation and so can be taken as evidence that the data are not adequately fitted by the model. That is, it shows that the sought-after interaction is present and hence is evidence of the presence of the difference in treatment responses.

In order to design a powerful experiment the following simulation experiment was performed.

1. Random samples from the multinomial distribution with means (0.30, 0.40, 0.20, 0.10) for treatment 1 and (0.00, 0.50, 0.30, 0.20) for treatment 2 were generated. Each sample was for a fixed number of experimental units (animals) in each group. Generating such samples is an easy standard operation available in many computer packages.
2. The parameters of a model not involving interaction between the classification and treatments of the contingency table were estimated using a generalized linear model.
3. Steps 1 and 2 were repeated 100 times and the number of times the critical value of 7.81 was exceeded was counted.
4. Steps 1 to 3 were repeated with other fixed numbers of experimental units.

Table 4.2.2 was obtained from simulated data using various sample numbers of experimental units in each treatment group. This table enables a sensible design strategy to be devised, taking account of the financial and ethical costs involved in running such an experiment. For

Table 4.2.2 The power of the experiment in Example 4.2.1

Number of experimental units in each treatment group	Number of times residual deviance is > 7.81 in 100 simulated data sets
15	69
20	79
25	92
30	97
35	100

many people, less than 20 animals in each group would result in an unacceptable probability of not detecting the difference. More than 35 animals in each group would result in an unacceptable ethical and financial cost as the probability of detecting the required difference is not increased with more animals.

The problem of fitting functions to time sequences of repeated measures on the same experimental unit frequently arises. This is because the data from each experimental unit are correlated from time point to time point. If the correlation is not accounted for in the fitting procedure, the standard errors of parameter estimates can be badly under- or over-estimated. When obtaining power functions by either analytical methods or simulation in this situation, the correlation between succeeding measurements needs to be known or, as above, a plausible guess substituted in the place of a known value. Parameter estimates and their standard errors can then be made using weighted least squares, as discussed in section 3.3. The nature of the correlations between time points can usually be obtained by judicious literature searches or informed guesses.

4.3 HOW TO MAKE COMPARISONS

Once a comparative experiment has been performed it is interpreted by comparing the parameter estimates using the standard errors of the differences in parameters. The standard error of the differences can be used to obtain a confidence interval from which a significance test is performed, as was done in the previous section. It is important to understand that a significant result does not mean the basis of the power assessment in the design phase of the experiment was correct. There is no way of ever knowing if this was true. After the completion of an experiment all that is known is what was known before the experiment was performed. That is, the probability of finding the specified difference using a significance test, had the basis of the power calculation been true.

Significance tests are used as part of the acceptance criteria for publication by journals and so, by using power as a design strategy and significance tests as an analysis strategy, an objective criterion for establishing and reporting worthwhile results is established. The reporting of only those results which are significant has the desirable effect of limiting the number of claims made in the scientific press which are spurious or about which there is too much uncertainty. There is an apparent disadvantage in not reporting uncertain results which, with hindsight, are seen to be worth reporting. This disadvantage can be minimized by designing an experiment which has sufficient power to find a response thought to be large enough to establish a case for suggesting (or not suggesting) alternative models of the phenomenon being studied.

The size of the confidence interval used in significance testing is not universally agreed. In fact it is a very thorny and controversial topic and is likely to remain so for as long as statisticians are able to argue. In the tests employed in section 4.2 above, each comparison is made with a 5% chance of finding a significant result when the true underlying means are the same. For an experiment with t treatments in which the t true means are all the same, there are $t(t-1)/2$ comparisons between the means. Therefore t does not have to be very large before it is highly probable that a type 1 error will occur. This is not acceptable to some people, particularly those who see significance tests as a means of categorically accepting or rejecting alternative models. Consequently many alternative strategies for significance testing have been proposed. If it is thought desirable to pursue such strategies, see for example Miller (1981, Chapter 2). However, regardless of an individual's judgement on this question, there is no reason why these alternative strategies cannot be used as a basis for the power calculations which have been used in section 4.2.

As an aid to interpreting rather than reporting the results of an experiment after it has been performed, the confidence interval is far more useful than a significance test. Yates (1964) highlighted the main problem of present-day analysis of experimental results:

> *The most commonly occurring weakness in the application of Fisherian methods is, I think, undue emphasis on tests of significance, and failure to recognise that in many types of experimental work estimates of the treatment effects, together with estimates of the errors to which they are subject, are the quantities of primary interest. In many experiments, e.g., variety and fertilizer trials, it is known that the null hypothesis customarily tested, i.e., that the treatments produce no effects, is certainly untrue; such experiments are in fact undertaken with the different purpose of assessing the magnitude of the effects.*

The confidence interval includes, with a specified probability, the unobservable difference in the true or real population parameters of the models which describe the response to treatments. Feasible alternative models can be suggested by considering this interval together with other already known information. The importance of other available information, some of it coming from the same experiment, should not be underestimated. Alternative models suggested in this way should then be assessed using information from a new experiment.

EXAMPLE 4.3.1

It was required to assess the ratio of the hazard, that is the proportional hazard, posed by a particular disease, malignant catarrhal fever, to hybrid deer with respect to the one susceptible species from which the hybrids were generated. The other parental species had a low susceptibility to the disease. The proportional hazard was estimated from the measured proportion of animals of each species dying from the disease each month. The proportional hazard of the hybrid animals with respect to the susceptible pure-bred parent was estimated to be 0.53, with a 95% confidence interval 0.25 to 1.09.

Because there was a 95% probability that 1 was one of the infinite number of possibilities for the true value of the proportional hazard, it was concluded that there was no significant difference between the hazards to the hybrids and pure-bred deer. Circumstances dictated it was not possible to conduct a further experiment in order to make a more powerful comparison with a narrower confidence interval. This was because one of the parent lines of deer had died out and so it was impossible to obtain any more of that species, or of the hybrids. Consequently, there was no alternative but to make the best use of the available estimate of the proportional hazard.

The confidence interval indicated to the investigator all the other values which might reasonably be the true value. This included the value 0.5, which is the value expected if the susceptibility to the disease was controlled by a single gene according to Mendel's theory. This possibility was of considerable interest in understanding the evolution of deer and a search was undertaken for corroborating evidence concerning the suggested genetics of the disease and its role in evolution of deer which competed with other grazing species.

4.4 PRECISION OF COMPARISONS

The precision of comparisons depends on the size of their standard errors. The standard errors are dependent on the size of the uncontrolled experimental variation. Two of the components of uncontrolled variation are imprecision in applying treatments and imprecision in measuring responses. Therefore, there is an opportunity to improve precision of comparisons by attention to experimental methodology. In many situations, attention to reducing experimental variation can be the most productive way of improving experimental efficiency. Careless experimental technique is a major source of uncontrolled experimental variation. The philosophy of total quality management (section 6.2) is enthusiastically recommended as an effective way of improving the quality of data. It has been recognized in manufacturing industries and some service industries, where it has been instrumental in yielding substantial rewards. A philosophy of trying to avoid introducing variation is much more sure of success than attempting to provide a list of rules which can never be complete, and for which compliance is often compromised by lack of commitment. Careful continuous appraisal of experimental techniques by everyone in the team, and adopting recommended improvements, is a satisfying and rewarding approach.

An incomplete list of experimental details, which can be fruitfully investigated in order to reduce uncontrolled experimental variation and hence improve precision, follows. By carefully considering each operation in a proposed experiment the list can be substantially enlarged.

4.4.1 CARE IN APPLYING TREATMENTS ACCURATELY

EXAMPLE 4.4.1

In an experiment designed to investigate the dynamics of salicylate distribution and excretion in deer, it was required to administer an intravenous dose of the drug. The dose was dependent on the weight of the animal. In order to minimize the uncontrolled experimental variation in the response from administering the drug, the weight of the animal had to be determined (not guessed), the volume of drug measured in a syringe with gradations small enough to ensure minimum error (preferably less than 1%) and the dose had to be administered by an operator expert enough to be sure the dose was delivered into the correct blood vessel. Clearly, a great deal of variation could be introduced into the responses of deer in this situation by quite large, easily avoided imprecisions in applying the experimental treatment.

4.4.2 CARE WITH ACCURACY OF MEASUREMENTS

There are many ways in which inaccuracies are introduced into measurements of experimental responses and it is important to assess each situation carefully in order to identify and remove the source of such inaccuracies. One important source is failure to ensure the free and unobstructed movement of weighing instruments.

4.4.3 ALLOWING FOR NUISANCE ENVIRONMENTAL EFFECTS

It is assumed that edge effects, or interference between adjacent plots, are unwanted nuisance effects. It could be that there was real interest in how the adjacent treatments interfered with each other, in which case different strategies might be more sensible. These effects can be due to gaps between plots where plants growing on the edge of the plots are often more vigorous and productive. Other interference can be because of the movement of fertilizers or from competition for nutrients, space and light.

In order to design away these problems in wheat trials, Lynch (1966) advised on a suitable field technique. The technique involved harvesting only the middle of the plots. There have also been techniques recommended for the analysis of data suffering from interference problems. However, if possible, avoidance of the problem is recommended as the most sure strategy.

4.4.4 ACCOUNTING FOR RESIDUAL EFFECTS FROM PREVIOUS EXPERIMENTS

EXAMPLE 4.4.2

Glass houses are often variable in response from side to side and also from end to end. In one such situation the practice was to lay out experiments in Latin squares. That is, each treatment occurred once in each row and once in each column. One possible layout for five treatments, A, B, C, D and E, is the following.

		Columns				
		1	2	3	4	5
R	1	B	C	D	E	A
o	2	A	B	C	D	E
w	3	D	E	A	B	C
s	4	C	D	E	A	B
	5	E	A	B	C	D

It was required to do experiments each year and, although a good deal of effort went into minimizing the effect of one year's treatments on the results of the next year, a further design strategy was used. It used the properties of Greco-Latin squares, which are given in Cochran and Cox (1957). The second year's design is given in Greek letters and is superimposed on the first year's design.

		Columns				
		1	2	3	4	5
R	1	Bβ	Cδ	Dα	Eγ	Aε
o	2	Aα	Bγ	Cε	Dβ	Eδ
w	3	Dδ	Eα	Aγ	Bε	Cβ
s	4	Cγ	Dε	Eβ	Aδ	Bα
	5	Eε	Aβ	Bδ	Cα	Dγ

Notice that each Greek letter occurs in each row and each column exactly once. Also notice that each Greek letter occurs with each Latin letter exactly once. Therefore, if there is a carry-over effect from the previous year, it occurs equally for each of this year's treatments. For a 5 × 5 Latin square this process can be carried on for a further two years before the original design has to be used again. The properties of Latin squares are discussed in Kempthorne (1952).

As well as using the properties of Latin squares, carry-over effects can be allowed for by the use of covariates, which are described in section 4.5.

4.4.5 ALLOWING FOR PLOT SIZE

EXAMPLE 4.4.3

Smith (1938) found in general that if σ_1^2 was the variance of plots of unit size, the variance of plots of size X was σ_1^2/X^b, where $0 < b < 1$. That is, there is always some advantage in taking a bigger plot size. However, the extra work in taking a bigger plot size has to be weighed against the increase in precision which can be obtained by increasing the number of replicate plots. In almost every situation it is better to increase the number of replicates than to increase the plot size. The increased plot size can only be used to advantage when there is relatively little extra work involved in its preparation and assessment.

4.4.6 SIMPLE OPERATIONS LEAD TO FEWER ERRORS

Complicated and hastily delivered and confusing instructions for laying down experiments and for observing responses are an important source of errors.

EXAMPLE 4.4.4

There is a class of experiments called resolvable designs. A resolvable design has a physical layout whereby one physical block of land constitutes one replicate of an experiment. It has been argued by many authors (for example, John and Williams 1997) that the use of these designs is advantageous because their simple layout results in a decreased number of laying-down mistakes.

4.4.7 USING TRAINED OPERATORS

EXAMPLE 4.4.5

In an experiment involving comparisons between weight gains of a number of breeds of cattle, two breeds were much more variable than the others. Furthermore, the expected difference between the two breeds was not as large as had been anticipated. By checking the ear tags with their nominated breeds, it was discovered that the data from the breeds had been mixed up. Very early on in the experiment an assistant had reclassified, incorrectly as it turned out, the animals on the basis of his understanding of how they should have looked. Perhaps the operator should have been trained not to tamper with experimental layouts; he should also have been better trained to recognize the breeds of the animals he was working with.

4.4.8 CAREFULLY DEFINING BOUNDS WHEN CLASSIFYING MEASUREMENTS

Precision of comparisons with categorical data can be improved by careful attention to defining categories and standardizing measurers. For example, tuberculosis lesion scores, bloat scores in cattle and liver damage scores. This can be achieved by training programmes at which all potential measurers are present. The basis of the training should involve demonstrations and detailed discussions as to what constitutes category members and category boundaries. This should be followed by practice in allocating unknown members to categories. This process should continue for as long as is needed to achieve the required standardization. The process should involve the writing of a more precise definition of categories and their boundaries, particularly if the training process is not achieving the desired results.

EXAMPLE 4.4.6

The tuberculosis lesion categories in the following table require careful definition and training to achieve a satisfactory result.

	Tuberculosis Lesion Categories			
	None	Mild	Moderate	Severe
Control				
Preparation A				
Preparation B				
Preparation A+B				

4.4.9 AVOIDING MEASURER BIAS

EXAMPLE 4.4.7

Double blind trials were devised to avoid measurer bias. They are a feature of medical investigations. In these investigations the people who have received an experimental treatment and the assessors of the response resulting from the treatment do not know what treatment has been given. This strategy overcomes the possibility that the subjects and the assessors may consciously or unconsciously influence the measured responses as a result of what they already know or would prefer to demonstrate. Even if none of these influences is operating, the double blind strategy disarms suggestions that they may have been. The need for methodologies to overcome bias in medical trials is clear. However, there is a strong case for applying similar methodologies in other experimental situations, including agricultural trials, where measurer bias could alter the perception of a response.

Measurer bias can also be introduced when a field plot or group of animals is sub-sampled. This is a complex problem and is treated in section 5.4.

4.5 COVARIATES

The previous sections have considered variates which will be modelled and about which inference is to be made. There is an additional and important class of variates which will now be considered. These are variates which coexist with the variates of primary interest. They are important because they allow potential sources of variation in the data to be allowed for in an analysis and subsequent interpretation. There are three ways in which these so-called covariates can be used.

First, covariates can be used as a background piece of information which can be considered when interpreting data.

EXAMPLE 4.5.1

During the seed planting season in the New Zealand high country there are periodic strong drying winds from the west. Should one of these occur as seeds germinate the results can be devastating as many seedlings wither and die through dehydration. The results from experiments in the high country are therefore very conditional on whether or not one of these winds occurs during the critical phase of plant establishment. This effect, however, is not the same for all species of plants, some being more resistant than others. It is very important to know when one of these strong winds occurred, as without this information the interpretation of an experiment is misleading.

Second, covariates can be used to adjust the estimated parameters of a model for an effect over which the experimenter has no control.

Example 4.5.2

A famous example of data where the estimated treatment parameters $\hat{\tau}_i$ of a simple additive model were adjusted for a single covariate Z, which has a different mean $\bar{Z}_i$ for each treatment, is presented in Figure 4.5.1. The data came from a large experiment on drugs used to treat leprosy which was conducted at the Eversley Childs Sanatorium in the Philippines. Leprosy bacilli were counted at six sites on the body of each patient and the counts scored both before and after treatment. Two of the treatments were experimental drugs and the third was a control. The relationship between the response, which is the scores after treatment, and the covariate, the pre-treatment scores is modelled by the parallel upward-sloping lines shown in Figure 4.5.1. Note how the lines for the two drug treatments almost coincide and are below the line which models scores in the control treatment.

The mean pre-treatment score for each of the treatment groups was 9.3, 10.0 and 12.9, and it is reasonable to ask if it was possible for the corresponding post-treatment responses, namely 5.3, 6.1 and 12.3, to

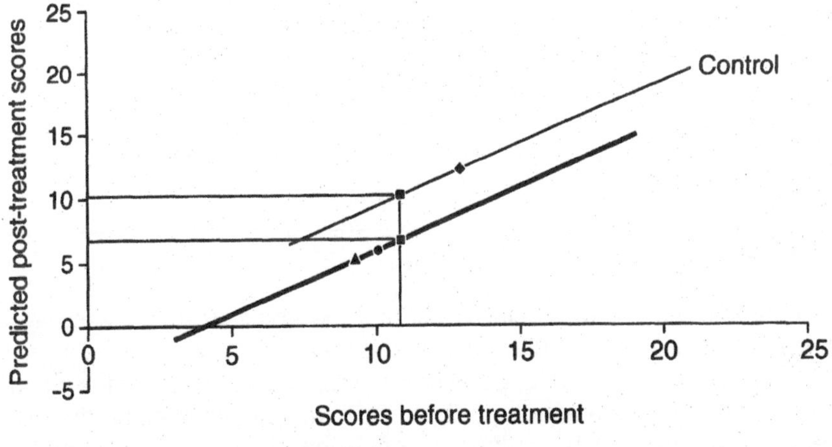

Figure 4.5.1 Fitted relationship between scores for leprosy bacilli before and after treatment.

be due simply to this difference and not to the applied treatments. The points (9.3, 5.3), (10.0, 6.1) and (12.9, 12.3) are represented on Figure 4.5.1 as a triangle, a dot and a diamond, respectively. Because of the strong relationship between the pre- and post-treatment scores within each treatment group, the effect of the pre-treatment scores can be allowed for by making the comparison between the treatments at the grand mean, 10.73, of the pre-treatment scores. This allows for the effect of the differing pre-treatment scores. The vertical line from the pre treatment score of 10.733 intersects the sloping lines representing the relationship between the pre- and post-treatments at 6.71, 6.82 and 10.16. These are the adjusted estimates of the treatment responses and they provide a more sensible comparison between the treatments than the unadjusted responses. However, the comparison between the adjusted responses has a larger standard error (1.846) and hence a wider confidence interval than at the means of the covariate for each of the three treatments (1.792).

Thus the gains in precision which resulted in modelling part of the uncontrolled experimental variation as $\hat{\beta}(Z - \bar{Z})$ can be partially or even wholly lost because of the loss in **efficiency** which results from adjusting for the modelled relationship.

More formally, this formulation may be expressed as a model as:

$$Y = \tau_i + \beta(Z - \bar{Z}) + \varepsilon,$$

where τ_i is the additive real or true response to the ith treatment, β is the slope of the response which depends on the covariate Z, $(Z - \bar{Z})$ is the difference between the covariate and its grand mean, and e is the uncontrolled variation in the measured response Y.

A covariate of this type which has attracted recent interest uses information from adjacent plots to adjust yields. One early and famous advocate of this technique was Papadakis (1937). His technique corrects the mean yields of individual plots by using the mean of the residuals of adjacent plots as covariates.

Finally, covariates can be used to improve the precision of the experiment.

Example 4.5.3

In Example 4.5.2 it was pointed out that the adjusted responses had a larger standard error (1.846) than at the mean of the covariate for each of the treatments (1.792). Sometimes, responses of replicates of the same treatment have a strong relationship with a covariate and the existence of the relationship is known before the allocation of treatments – that is, in the design stage of the experiment. An example is the relationship between an animal's live weight gain over a specified time interval and its weight at the start of the interval. Amongst other possibilities, the weight gain can also be related to its known birth rank, rearing rank, gender and sire.

In order to show how covariates can be used as an aid to designing experiments with improved precision, consider the situation where there is a single covariate which has the same mean, the grand mean, for each treatment. An example is presented in Figure 4.5.2. If the relationship between the response and the covariate is modelled by parallel lines as shown in Figure 4.5.2, a substantial improvement is obtained in the precision of comparisons between treatments. This is because a proportion of the otherwise uncontrolled experimental variation is modelled by estimates of the equation represented by the parallel lines – that is, $\hat{\beta}(Z - \bar{Z})$. In Figure 4.5.2 note that the mean of the covariate for the two treatment responses is the same, and so the fitting of the parallel lines has not carried with it the opportunity to adjust the comparison between the two treatment responses with a potential loss of efficiency.

The example presented in Figure 4.5.2 shows that there is a clear advantage in having a similar distribution of experimental units for each of the treatments when covariates are present and are allowed for. This result generalizes for when there are many treatments, and for when there are many covariates. The desirable similarity between the distribution of experimental units within an experiment can be achieved as follows. The experimental units are arranged into groups which are as uniform as possible as far as the covariate is concerned. The size of each

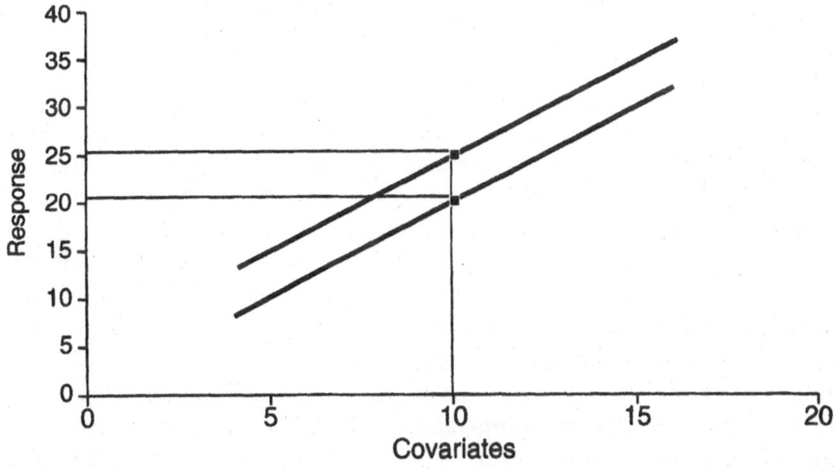

Figure 4.5.2 Fitted relationship between a response and a covariate.

group is the same as the number of experimental treatments. The experimental units from each group are then allocated to treatments at random, one experimental unit from each group to each treatment.

EXAMPLE 4.5.4

One of three treatments was to be applied to pregnant mares, and the rate of growth of their foals was to be compared. In all, 54 mares provided 18 replicates of each treatment. The age of each of the mares and the sires, of which there were two, were covariates which were to be allowed for in the design.

Within each sire group the mares were sorted into ages and allocated consecutively into groups of three. The three mares in each group were then allocated at random, one to each of the treatment groups. The final allocation was as shown in Table 4.5.1.

The range of the mares' ages and the use of both sires within each treatment allowed for any within-treatment relationships between these covariates and the growth response to be modelled. This was intended to result in improved precision of the comparisons. The balancing of the ages and sires between the three treatments ensured that any adjustments in the treatment comparisons were small and so the improved precision was not compromised.

Covariates provide a way of comparing profiles of repeated measurements where no specific features of the profiles are of interest before the start of an experiment. Details are given in Kenward (1987). The basis of the method is to establish the number r of preceding repeated measurements beyond which the current measurement is independent. The estimated treatment responses in the current variate are then adjusted for the immediately preceding r variates which are treated as covariates. The adjusted estimated treatment responses are then compared using a significance test. The advantage of doing this rather than comparing the unadjusted estimated responses of each treatment at each design point is that the tests using the adjusted estimated responses are independent of each other. Each time a significant difference occurs it may be taken as an indication that the profiles have changed shape at that point. The first test in a sequence involves no covariance adjustment and is less sensitive than subsequent tests as it is the only test not involving covariance adjustment. If the profiles have a large number of design points, the number of significance tests which are performed becomes large, a problem which was referred to in section 4.3. An application of the method from Kenward (1987) is presented in the following example.

Table 4.5.1 Application of treatments to mares in Example 4.5.4

		Treatment			
Sire	Age	1	2	3	Number
Account	18	1	0	0	1
	17	0	1	1	2
	16	1	0	1	2
	15	0	0	0	0
	14	1	1	1	3
	13	0	1	0	1
	12	0	1	0	1
	11	2	0	1	3
	10	1	2	1	4
	9	0	0	1	1
	8	1	1	0	2
	7	0	0	2	2
	6	1	1	0	2
	5	0	0	0	0
	Number	8	8	8	24
Runaway	18	0	0	0	0
	17	0	0	0	0
	16	0	0	1	1
	15	1	2	0	3
	14	0	0	0	0
	13	0	0	0	0
	12	1	0	2	3
	11	1	2	1	4
	10	3	1	1	5
	9	0	1	0	1
	8	2	1	2	5
	7	1	2	1	4
	6	1	0	2	3
	5	0	1	0	1
	Number	10	10	10	30

EXAMPLE 4.5.5

It was required to compare the effect of two treatments, nominally called A and B, on the growth of 30 cattle subjected to a pasture parasite challenge during the course of a growing season. During the season the weight of each animal was recorded 11 times to the nearest kilogram. The weights were separated by two-week intervals except the last, which was only one week after the previous weighing. The profiles for the average weight of the two treatment groups were as shown in Figure

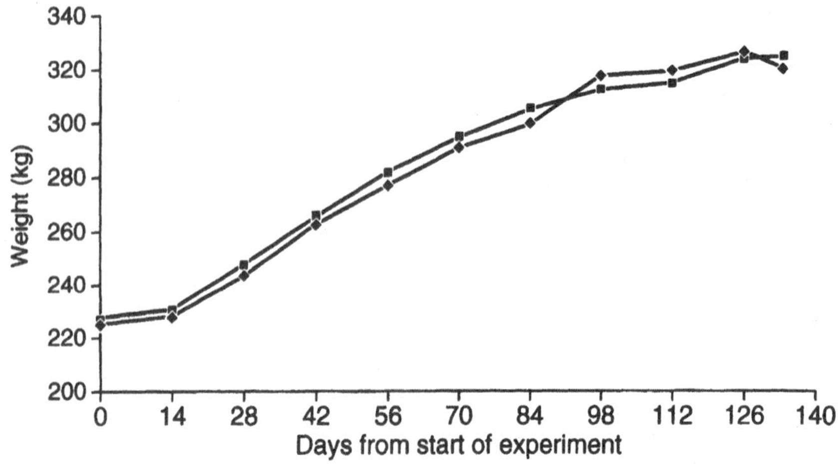

Figure 4.5.3 Weight plotted against days from the start of the experiment.

4.5.3. The average responses which are presented in Figure 4.5.3 were adjusted using the previous two measurements on the individual animals for covariance adjustment. The adjusted responses on day 98 and day 135 were significantly different, the points at which the unadjusted profiles cross over. This result focuses attention on those two dates as being important when attempting to understand and model the differences between the two treatments.

4.6 CONFOUNDING

Confounding is the perplexing situation presented when there are a number of plausible explanations for an observation. When designing an experiment it is best to devote some time to considering what the confounding influences might be. There is a clear advantage in being able to eliminate them or minimize their effect.

Confounding occurs if the experimental treatments are segregated in either time or space. An example of segregation in time is when a field experiment has all replicates of one treatment planted on one day and all replicates of another treatment planted on a different day. When it comes to interpreting the result the consequences of the segregated planting will not be separable from the treatment effects.

Comparisons between measurements which are spatially separated offer similar opportunities for alternative explanations. As an example, differences between soil types throughout a country are confounded with all the other geographical differences between the regions.

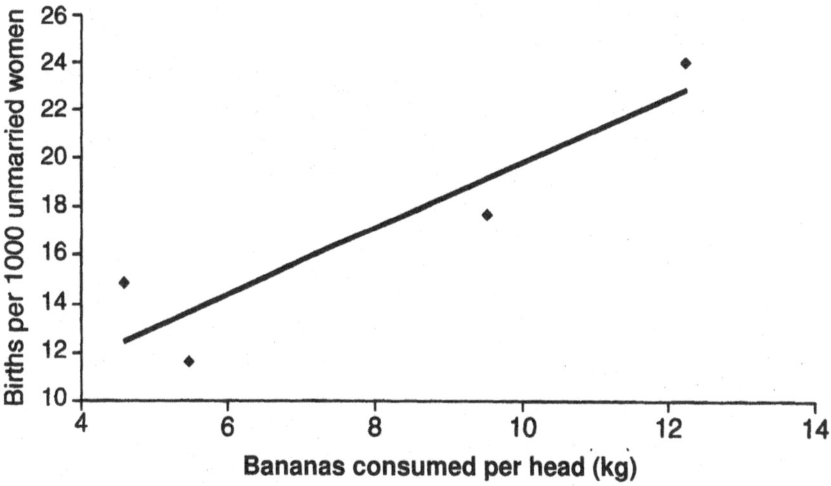

Figure 4.6.1 A nonsense relationship.

Consequently, care is needed when interpreting differences in terms of different soil types which are in different geographical regions.

The most common confounding effects are when measurements are changes in time. There are many famous examples of misinterpretation of such results. To illustrate one such possibility, Yule (1926) drew attention to the high correlation between membership of the Church of England and suicide rate. It was because both variates, membership of the Church of England and the suicide rate, were changing at the same time. Other examples are easy to find.

Figure 4.6.1 displays another of these so-called nonsense correlations, between the consumption of bananas per head and the birth rate amongst unmarried women in New Zealand between 1945 and 1961. Although it is easy to point to such obvious examples it is also very easy to make misinterpretations, when the confounding variate offers a more plausible explanation. Opinions expressed in newspaper reports are a fruitful source of such 'explanations'.

Problems associated with interpreting changes in time can be minimized by careful design strategies. Experimental units which are left alone allow for the assessment of changes that would have taken place if no treatment was applied. However, the control groups have to be typical of the treated groups. This is achieved by randomizing (dealt with more fully in section 5.3) all experimental units into nominated treatment and control groups. The chance of an allocation of atypical experimental units to any one treatment or control group is thus minimized, but not eliminated. This small risk can either be accepted, or other covariates which are thought to affect the outcome can be identified and subsequently used to help interpret the data. It may be tempting to try to even up the random allocation if it is found to be wanting. However, it is better to identify disparate experimental units before the random allocation is undertaken and use the strategy of blocking (which is discussed in section 5.2) or covariance analysis (which was discussed above in section 4.5), to help eliminate the unwanted effect.

EXAMPLE 4.6.1

It was required to assess the potential untoward effects on dogs of a single dose of the drug praziquantal, which eliminates all species of tapeworms in dogs. Potential acute toxicity was assessed by allocating at random 40 dogs into four groups of ten. One of the groups was assigned to receive no drug and the other three groups were treated with a single dose of the drug at the recommended dose, five times the recommended dose and ten times the recommended dose. Blood samples were taken at weekly intervals for three weeks before dosing and at 12, 24, 48 and 168 hours after treatment. In all, 12 blood measurements were monitored. The pre-dosing measurements were used as a covariate in order to model and remove the effect of individual dogs. The control group allowed the sensible interpretation of changes in the treated dogs after they had been dosed. Without the control group, changes in blood measurements which were associated with the environment may have been misinterpreted as being due to the actions of the drug.

An example of a design which is used to eliminate both spatial and temporal effects comes from environmental impact assessment.

EXAMPLE 4.6.2

Environmental impact assessment is difficult because such impacts are confounded both spatially and temporally. For example, changes in a specific area after an oil spill may be due to influences other than the spill which impact all associated areas, and changes in the area over time may have occurred because of some natural fluctuation which is independent of the spill. Consequently, so-called Before After/Control Impact (BACI) designs have been proposed to overcome these difficulties. These designs supposed that it is known that a potential impact is imminent, such as the building of a thermal power station which discharges warm water into a river or the conversion of farm land to alternative farm types. The idea is to identify reference areas which are similar to areas where the environmental impact is to occur and to make measurements at all sites both before and after the impact takes place.

The importance of replicating the reference area has been emphasized by Underwood (1994). The need for replication is further discussed in section 5.1.

4.7 REDUCING COSTS

The costs of conducting an experiment do not necessarily increase uniformly as the effort increases. This is particularly so if the laying down of the experiment or an assessment involves travel. In this case there is often a small difference between the cost of an experiment which is the maximum size which could be serviced in a day and a smaller experiment involving only part of a day. The criterion of doing experiments which occupy multiples of a full day's work in laying down and assessing is a good one when used in conjunction with a power assessment.

Another strategy for reducing costs is sub-sampling of harvested material. An example is the weighing of 1000 grains of a cereal rather than counting all the grains from a harvested plot. The cost is increased variation but this may not be very important as it is often very small. Suppose the variation amongst the sub-samples is σ_1^2 and between the plots is σ_2^2. If there are N sub-samples in a plot the variance of plot means is

$$\sigma_2^2 + \frac{\sigma_1^2}{N}.$$

If only n sub-samples are taken the variance of a plot mean, calculated from the n sub-samples, is

$$\sigma_2^2 + \frac{\sigma_1^2}{n}.$$

The increase in variation resulting from taking the smaller sample is given by the difference between the two equations,

$$\frac{\sigma_1^2}{n} - \frac{\sigma_1^2}{N},$$

and this is often small when compared with what it would be if all the plots were considered as σ_1^2 is often small when compared with σ_2^2. This increase in variation may be judged to be a worthwhile penalty to pay for the decreased effort and financial savings resulting from a less intensive sampling effort.

As an example consider the situation when $\sigma_1^2 = 10$, $\sigma_2^2 = 50$, $N = 1000$ and $n = 10$. If the $N = 1000$ sub-samples are taken the variation of plot means, the variance, is 50.01. If only $n = 10$ sub-samples are taken the variance is 51.00.

As with increasing the plot size, if interest is in estimates of the amount of harvested material, it is far better to increase the number of replicates than increase the intensity of sampling from the replicates. If interest is in within-plot variation a more intense sub-sampling effort may be indicated.

Experimental layout 5

5.1 REPLICATION

When experimental measurements include a component of uncontrolled variation the measurements on replicate experimental units are recognized as the basic requirement enabling a confident interpretation of the experiment. Increasing the number of replicate experimental units increases the power of comparisons, as has been discussed in section 4.2. At the same time replication improves the precision of estimated treatment differences by reducing the size of the standard errors of the estimated model parameters.

Generally, the size of standard errors of parameter estimates or the size of the standard errors of differences in parameter estimates is proportional to the inverse of the square root of the number of independent observations on independent experimental units. Consequently, quadrupling the number of replicates of an experiment halves the size of the standard errors or standard errors of differences. Therefore, increasing the number of experimental replicates is used effectively to complement the judicious choice of design points (discussed in section 3.3) and the minimization of uncontrolled experimental variation (discussed in sections 4.4 and 4.5).

There are other reasons for replicating experiments. Replication of experiments in time and space enables temporal and spatial variability of responses to be estimated. Physicists who can do definitive one-off experiments, with effectively no uncontrolled variation obscuring the treatment responses, like to have their results reproduced by replicate experiments, preferably in other laboratories. This excludes the possibility that some unaccounted within-laboratory effect or coincidental time effect has unduly influenced the measurements.

In some situations replication is impossible. This may be because of the size of the experimental unit being studied, such as a bird colony, or the cost. It may also be because the whole population consists of just one experimental unit! In this situation there is no reason why unreplicated experiments cannot provide valuable information. A single case study is far more value than no information, particularly if large responses are expected. Unfortunately, with such experiments there is a tendency to pretend that measurements made on different parts of the same experimental unit constitute proper replication. When this is practised the replicated measurements are really pseudo-replication because the observation does not include an independent component of uncontrolled variation. An awareness and understanding of pseudo-replication is important as it can be unwittingly practised.

Pseudo-replication is a phenomenon whereby the observations which are taken as replicates do not necessarily reflect the variation in the population of experimental units about which information is being sought. As has been said, it occurs when the replicated observations are not independent of each other. It can occur in experiments involving field crops or other field plants such as weeds by applying a particular treatment to a continuous part of the field in one sweep and taking measurements from the treated area and considering them as if they were from properly replicated plots. Such a practice ignores the variation induced in observations by inaccuracies in independently applying the treatment to the individual plots and the effect of spatial variation over all the field. Consequently the uncontrolled variation amongst the pseudo-replicates is likely to be smaller than the variation which would have been observed with independent replicates. It is therefore unwise to rely on standard errors which are estimated from pseudo-replicates instead of independent replicates as they will be biased downwards.

Another example of pseudo-replication is when the experimental unit is a group of animals, and individuals within the group are treated as replicates. This is attractive as the size and cost of the experiment can be substantially reduced if this strategy can be used. Conniffe (1976) used some data from some Irish grazing trials in which groups of animals were the experimental unit. He was able to demonstrate that valid conclusions could be drawn from an analysis in which individual animals were considered to be replicates. In the situation being considered this was possible because the between-group variation was small when compared with the within-group variation. However, it should be noted that this observation could be made only after the properly replicated experiment had been completed. The size of the variation between experimental units cannot be known prior to the conduct of an experiment, and it is therefore unwise to rely on pseudo-replicates to provide an unbiased estimate of uncontrolled experimental variation on the experimental unit, the group of animals, when designing such an experiment.

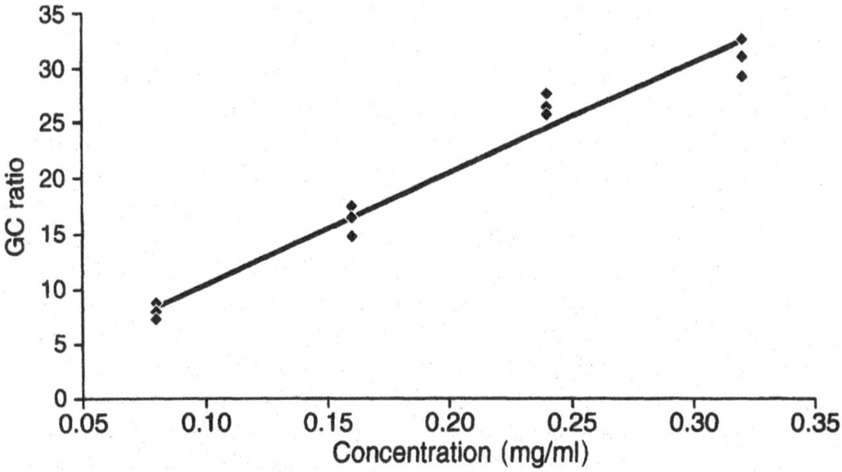

Figure 5.1.1 A calibration curve.

Calibration of an analytical instrument provides another opportunity for pseudo-replication. If a single dilution is repeatedly measured instead of a fresh dilution being made for each measurement, only the innate variability of the machine will be reflected in the different measurements and the mean of the measurements will include a common bias caused by inaccuracy in making the single dilution. The calibration curve displayed in Figure 5.1.1 was prepared from data from three replicate determinations of four single dilutions. The variation at each of the design points reflects only the innate variability of the machine, and the lack of fit from the straight line expected in the calibration curve may be due to inaccuracies in preparing the single dilution used at each design point. The preparation of a new dilution for each measurement would result in a measure which included the variation induced by inaccuracies in making the dilution. These would tend to cancel out when the mean measurement was calculated and so eliminate the effect of the bias in the measurements when just a single dilution was used.

A further example is the use of repeated observations of the same experimental unit as if they constituted an independent set of observations, that is, proper replicates. For example, daily observations on a single ecological entity, such as the bird colony mentioned above, do not constitute an independent set of observations. Hurlbert (1984) has drawn attention to the prevalence of this in ecological studies and detailed many examples.

It was pointed out in section 3.3 that when the treatments are structured and there are a large number of factors, it is sometimes possible to model the response with interaction terms limited to those involving three or, in many situations, even less factors. Estimates of parameters for these reduced models can be made from a judicious choice of a fraction of a single replicate of the complete experiment. However, the use of a single fractional part of a complete replicate of an experiment seems to ignore the need for replication. Fortunately this is not so.

In factorial experiments there is much built-in replication of treatments, as with each replicate of the experiment each treatment occurs once in combination with each other treatment. This means that in the 2^8 example discussed in section 3.2 there are 128 treatment combinations which include one of the two levels of each of the eight treatment factors. If there are enough treatment factors in the experiment, and in the 2^8 example there almost certainly would be, this built-in replication allows the numbers of replicates of the experiment to be reduced. As has been already noted, this can result in just a single replicate or even a fraction of a single replicate of the experiment being sufficiently powerful to enable the comparisons of interest to be made.

5.2 BLOCKING

Blocking involves the identification of uniform plots of experimental material which are called the blocks, and allocating the experimental treatments to the plots within the blocks. The variation between the blocks can then be modelled in the same way as any other factor which is part of the experiment. Consequently blocking enables a large part of the otherwise uncontrolled variation to be removed from the estimate of uncontrolled experimental variation. As the estimate of uncontrolled variation contributes directly to the size of the standard errors of the estimates of the model parameters and their differences, blocking reduces the size of the standard errors of treatment comparisons.

EXAMPLE 5.2.1

The precision of the experiment of Example 3.2.1, the measurements from which were modelled by

$$Y = \tau_i + \varepsilon,$$

may be considerably improved by modelling part of the uncontrolled variation ε, by a block effect. In this example it is reasonable to expect environmental differences between flocks within a farm to be less than environmental differences between farms. Consequently the effect of the farm may be modelled as blocks, thus reducing the estimate of uncontrolled experimental variation. This was in fact the planned strategy when this experiment was designed. The modelling of the farm effect results in the estimate of the standard error of the difference between the estimated treatment effects changing from 554.8 to 138.4, which is a substantial improvement.

Most computer packages which perform statistical analysis have the capability to model the block structure of an experiment. However, care is needed to specify the structure correctly to ensure the required modelling takes place. Failure to do so results in the loss of the enormous potential which has been demonstrated in Example 5.2.1.

When selecting blocks, the experimenter should have reason to believe that they are uniform. In an animal experiment this might be because members of a block all have the same sire, or are the same age, or are from the same environment. In field experiments blocks might be land of known soil type, depth and aspect. A block of land may not necessarily be continuous, and can be composed of a number of patches which are similar. The ideas of blocking can also be extended to include the effect of particular operators or days when, for example, treatment application involves the expenditure of much labour or time.

It is common in agricultural experiments to have blocks all the same size and the same number of replicates for all treatments. When this occurs many block designs are available. These designs can be either complete or incomplete. Complete block designs occur when the number of plots in a block is equal to the number of individual treatments. Incomplete block designs are those where the number of plots in a block is less than the number of individual treatments. Incomplete block designs may be balanced. This occurs when the number of plots available is equal to the product of the number of treatments and replicates, and the treatments occur together in the same block the same number of times.

An example of a balanced incomplete block design with seven blocks, each with four plots, and seven treatments is the following:

Block	Treatment			
1	1	2	5	7
2	3	5	6	7
3	1	2	3	6
4	1	4	6	7
5	2	4	5	6
6	1	3	4	5
7	2	3	4	7

Close inspection will reveal that each treatment is replicated four times and each pair of treatments occurs together in the same block twice. Partially balanced incomplete block designs occur when each pair of treatments is together in the same block almost the same number of times.

An example of a partially balanced incomplete block design follows. There are 12 blocks, each with three plots and 12 distinct treatments.

Block	Treatment		
1	1	2	3
2	4	5	6
3	7	8	9
4	10	11	12
5	4	7	10
6	1	8	11
7	2	5	12
8	3	6	9
9	6	8	12
10	1	5	7
11	2	9	10
12	3	4	11

There are 66 pairs of different treatments in this design and three pairs can occur in each block. In the first block the pairs are (1,2), (1,3) and (2,3). In the whole design there are 36 pairs which can occur together. As it turns out, those 36 pairs which do occur are all different pairs. Therefore the number of times treatments occur together in this design is either one or zero and the design is partially balanced.

Balanced incomplete block designs are not available for many treatment and replicate combinations. The balanced incomplete block designs and partially balanced incomplete block designs are catalogued in many books, details of which are given in Appendix A. A computer program, ALPHA+, commercially available from Biomathematics and Statistics Scotland, generates incomplete block designs with up to 500 treatments. GENSTAT, a computer program mentioned above, has an extensive collection of procedures which includes an ability to generate these designs.

Factorial experiments can have many more treatment combinations than the number of uniform plots which could be reasonably expected to be found in a block. There are strategies for confounding unwanted treatment comparisons with block comparisons when the available blocks of uniform plots are all the same size. Books with details of these designs are listed in Appendix A. GENSTAT and DSIGNX generate such designs.

As noted above, it is common in agricultural experimentation for the blocks to have the same number of plots. However, this is unnecessarily restrictive. Residual maximum likelihood (section 3.3) easily copes with blocks which have a variable number of plots. This allows more precise experiments as blocks of natural size can be used for treatment allocation rather than imposing the analytically convenient, but sometimes rather artificial, requirement for blocks of equal size. It is very unlikely that rams, when working at generating blocks of offspring, will have the potential experimenter in mind and, in consultation with peers, agree to service an equal number of ewes each. Even if they did, would the ewes have the same number of male and female lambs? It also seems unlikely that similar principles guided the laying down of agricultural land.

The design of an experiment involving unstructured treatments in blocks with unequal numbers of plots is not difficult.

EXAMPLE 5.2.2

Suppose five treatments are applied to four blocks in which there are 3, 6, 2 and 4 plots, respectively. If equally replicated treatments are allocated so that pairs of treatments occur together in the same block as close to an equal number of times as possible, an efficient design is obtained. That is, the average standard errors of the differences between treatments will be as small as possible. As a start to finding a suitable allocation it is worthwhile applying treatments sequentially to the available plots, as has been done below. If this initial design is not satisfactory, reallocation of treatments to a new design is easy.

Block	Treatment
1	1,2,3
2	4,5,1,2,3,4
3	5,1
4	2,3,4,5

In this example, suppose the effects of blocks are random samples from a normally distributed population with a mean of zero and a standard deviation of 5. Also suppose the uncontrolled experimental variations are samples from a normal distribution with a mean of zero and a standard deviation of 1. Then the average standard error of estimated treatment differences is about 0.85.

If the total number of plots is not a multiple of the number of treatments it is better to allocate all the plots to a treatment and have some treatments replicated one more time than others.

The design of experiments with a factorial structure, in which the blocks have unequal numbers of plots, is an important class of experiment. As with unstructured treatments, it offers the experimenter the ability to exploit blocks of natural size in order to increase precision of comparisons. A technique for allocating treatments to plots within unequal sized blocks is illustrated in the next example.

EXAMPLE 5.2.3

Suppose it is required to conduct a factorial experiment in which each of three levels of one factor is combined with each of two levels of another. That is, there are six unique treatment combinations. Exactly the same strategy as in Example 5.2.2 may be used to allocate treatments to plots. That is, the six treatment combinations are allocated to the available blocks so that the pairs of treatments occur together in the same block as close to an equal number of times as possible.

In some situations the comparisons amongst the treatments may not be all of equal importance or the number of plots may not be a multiple of the number of treatments. This is when the technique of scoring can be used to advantage.

Scoring is performed as follows. Treatment combinations are allocated to blocks. A random allocation is often a convenient starting point. Comparisons of interest are then scored within each block. This is done by counting the number of observations which are represented by each side of the comparison and comparing the counts.

EXAMPLE 5.2.4

The following example is taken from Table 10 of Mead (1990). The scores are calculated for the displayed allocation of treatments to three blocks. The blocks are of size 4, 5 and 6 plots. The treatments are for a 2 × 2 × 3 factorial structure which forms 12 individual treatment combinations. The factors are A, B and C and have levels designated as a_0 and a_1 for factor A, b_0 and b_1 for factor B, and c_0, c_1 and c_2 for factor C.

Block 1	Block 2	Block 3
$a_0\, b_0\, c_0$	$a_0\, b_0\, c_2$	$a_0\, b_0\, c_0$
$a_0\, b_1\, c_2$	$a_0\, b_1\, c_0$	$a_0\, b_0\, c_1$
$a_1\, b_0\, c_1$	$a_0\, b_1\, c_1$	$a_0\, b_1\, c_2$
$a_1\, b_1\, c_0$	$a_1\, b_0\, c_0$	$a_1\, b_0\, c_2$
	$a_1\, b_1\, c_2$	$a_1\, b_1\, c_0$
		$a_1\, b_1\, c_1$

Comparisons	Block 1	Block 2	Block 3	Final scores
$a : a_1 - a_0$	0	−1	0	0.20
$b : b_1 - b_0$	0	1	0	0.20
$c' : c_1 - c_0$	−1	−1	0	0.45
$c'' : c_0 + c_1 - 2c_2$	1	−1	0	0.45
$ab : a_1(b_1 - b_0) - a_0(b_1 - b_0)$	0	−1	2	0.87
$ac' : a_1(c_1 - c_0) - a_0(c_1 - c_0)$	1	−1	0	0.45
$ac'' : a_1(c_0 + c_1 - 2c_2) - a_0(c_0 + c_1 - 2c_2)$	3	−1	0	2.45
$bc' : b_1(c_1 - c_0) - b_0(c_1 - c_0)$	−1	1	0	0.45
$bc'' : b_1(c_0 + c_1 - 2c_2) - b_0(c_0 + c_1 - 2c_2)$	−3	1	0	2.45

The final comparison scores in the right-hand column of the table above are then calculated from the individual block scores as the sum of the square of the individual block scores divided by the number of plots in the block. The smaller the final score for the contrast, the smaller will be the standard error of that contrast.

The scores are then used as an indicator of where the reallocation of treatment combinations to blocks would improve the design by obtaining scores and hence standard errors which best suit the purpose in hand. Comparisons which are of little or no interest can be arranged within the blocks to have higher scores.

In some situations pieces of experimental material may have environmental gradients at right angles to each other. An example is given in Example 4.5.2. The Latin square in that example was used to eliminate two sources of variation operating at right angles to each other. It is often advantageous to eliminate two sources of variation operating at right angles when the very restrictive limitations imposed by a Latin square – that is, the numbers of rows, columns and treatments must all be the same – cannot be complied with. One strategy for dealing with a number of rows which differs from the number of columns is to truncate a Latin square at a suitable place. When this is done the Latin squares become a Youden square. However, the limitations on the structure of row column designs (as they are called) can be further lifted. There are computer algorithms which design efficient rectangular row column designs. One of these is incorporated into the packages ALPHA+ and GENSTAT. However, it is possible to obtain adequate designs by simple *ad hoc* procedures, as the following example demonstrates.

EXAMPLE 5.2.5

A field with seven columns of plots was available for an experiment. The columns had unequal numbers of rows, there being 17, 16, 16, 15, 14,

14 and 13 rows respectively in each column. It was required to allow for obvious row and column effects in the design of an experiment involving seven replicates of 15 treatments. Comparisons between pairs of treatments were all equally important. The 15 treatments were put into random order seven times, a different randomization being used on each occasion. The treatments were then allocated sequentially down the first column, into the second and so on till the end of the seventh column. The design, with treatment numbers entered in the main body of the table, was as follows:

	Column						
Row	1	2	3	4	5	6	7
1	1	13	10	6	5	10	9
2	7	1	11	5	14	4	11
3	12	6	7	7	13	12	8
4	8	11	8	3	11	15	6
5	9	15	2	11	12	13	3
6	3	2	6	15	10	8	2
7	5	10	4	14	15	3	7
8	14	5	5	1	4	5	14
9	2	12	1	9	3	11	13
10	4	9	13	8	7	1	12
11	15	7	14	2	1	9	1
12	13	4	9	2	2	6	10
13	6	8	13	9	7	15	5
14	10	12	12	6	14	4	
15	11	15	4	8			
16	14	3	10				
17	3						

For assumed known variation between the rows and columns and between plots within the rows and columns, it is possible to obtain the standard errors for each pairwise comparison using theoretical results which are presented in Patterson and Thompson (1971). The average and range of the standard errors of differences (or possibly other interesting comparisons) can then be calculated and judgements made concerning the adequacy of the design. If it had been required to narrow the range of standard errors of differences an attempt would have been made to equalize the number of times each pair of treatments occurred together in the same row and in the same column. This would have been done by randomly reallocating obviously excessive repetitions of the same treatment in any one row or column. In the above design, the occurrence of treatment 5 in row 8 would be a candidate for such reallocation. Alternatively, and perhaps more simply, another random allocation could be tried.

However, instead of using theoretical results, simulation is sometimes more convenient and in fact was used to assess the above design. In an analogous way to that described in section 3.3, simulated data were generated using plausible variation between the rows and columns and between plots within the rows and columns. This was subjected to model fitting using residual maximum likelihood where it was found that the standard error of the differences between pairs of treatment responses differed from the average by at most 10%. This was considered to be an adequate design.

There is a class of designs called cross-over designs. These are masquerading block designs. Cross-over designs occur when an experimental unit has a sequence of different treatments applied to it over a period of time. The experimental unit can be thought of as the block and the plots are the different periods in the time sequence. The advantage of these experiments is that comparisons between treatments are made on the same experimental unit. This eliminates variation between experimental units in treatment comparisons and so there is scope for much improved precision. The disadvantages are twofold. First, differences in responses to different treatments on the same experimental unit are confounded in time. Second, there has to be sufficient time between the application of treatments for the effect of one treatment to not influence the next treatment. This time interval between treatments is often called the wash-out period.

The first disadvantage can be allowed for by having all the treatments being applied to some of the experimental units at each time period. In a two-treatment experiment this would mean that the first treatment would precede the second treatment in half the experimental units and the second treatment would precede the first in the other half. This idea can be generalized to a three-treatment experiment. Each one-sixth of the experimental units would be given sequences of treatments as follows:

1,2
2,1
3,1
1,3
2,1
3,2

The second disadvantage can be allowed for by carefully modelling the carry-over effects. However, if possible, it is best to avoid ambiguities in interpretation of the experiment and so cross-over experiments are best used when the response to treatments can be assumed to be independent of the order of application of the treatments.

There are two further blocking techniques which are referred to by another name. The first is called paired comparisons. This involves applying two treatments which are to be compared, one to each of a matched pair of plots. This is in the spirit of blocking as the aim is to make comparisons between treatment responses on uniform experimental plots.

EXAMPLE 5.2.6

Suppose a treatment is allocated to one of a pair of twin lambs, the untreated twin being maintained as a control. The comparison between the twins eliminates a large component of variation from the comparison as the twins share a common environment. The block is the pair of lambs.

The second further blocking technique is called stratification. Stratification is a name used to describe treatment allocation when the experimental units are matched on some attribute (such as weight), and a different treatment applied to each of the matched individuals. The term is often used with animal experiments when the treatments are applied to the individual animals. If the differences between the strata are not modelled in the analysis the effect of stratification is to increase the magnitude of the estimate of uncontrolled experimental variation rather than decrease it as was intended.

EXAMPLE 5.2.7

It was required to compare the response to two forms of colostrum replacer given to orphaned new-born lambs. It was hoped the protein would increase the survival rate and improve the growth performance of such lambs. The experiment required two dose levels of one of the protein forms and three of the other. Lambs were required to be given the protein within a few hours of birth. Therefore, as new-born lambs became available they were given their treatment. After each group of five lambs had been treated, each with one of the five treatments, they were put together in a protected environment. Each stratum of lambs stayed together for the duration of the experiment. It would be expected that the uniform environment within each stratum of five treated animals would allow more precise comparisons within the strata than between the strata. The word 'strata' in this context is equivalent to 'block' and it is a matter of semantics as to what it is called.

The consequences of not blocking when it is needed or of blocking incorrectly can be serious. This is because a large component of experimental variation which could be modelled is assigned to the estimate of the uncontrolled experimental variation, thus increasing its magnitude.

EXAMPLE 5.2.8

In Example 5.2.2, if the need for blocks is ignored, the average of the standard errors of the differences between means is five times larger than if the blocks are allowed for. Incorrect blocking will result in standard errors somewhere between the two extremes, the actual position between the extremes being dependent on the degree of incorrectness when compared with the ideal.

Sometimes there is no apparent justification for blocking. This is because the experimental material seems to be uniform. However, invisible fertility gradients frequently occur in field experiments, and invisible temperature and lighting gradients occur in glass houses. Therefore it is good practice to block the experimental area into smaller pieces. A general rule is to block as long as there are more than 12 degrees of freedom for the estimate of the uncontrolled experimental variation. In this situation there is nothing to lose from blocking even if the blocks do not account for any of the uncontrolled variation. However, there is a chance that there will be a considerable gain by removing variation from an undetected gradient in fertility. Long narrow blocks in the face of no information about the within-blocks variation are best avoided. This is because of the possibility of large variation between the plots at the extreme ends of the block.

One possible alternative to blocking is to model the underlying uncontrolled variation of the experimental material. This has been done very successfully in wheat variety trials. Cullis and Gleeson (1989), in a study of over 1000 variety trials, demonstrated a reduction of 42% in the variance of varietal yield differences and compared this to the 33% achieved with incomplete block analyses. Details of the methodology are available in Gleeson and Cullis (1987) and Cullis and Gleeson (1991).

If there is more than one observer each doing independent assessments of trial plots, each observer should collect data from a complete block. This confounding of the observer and block differences results in observer differences being removed from treatment comparisons with block differences.

It is common practice for experimenters to conduct split-plot experiments. These are used when the size of the plot required for one set of treatments is different from the size of the plot required for another set. As an example, consider the size of the plots required for the adequate application of alternative reclamation techniques after open-cast mining. Because of the size of the machinery used, the plots have to be very large. Compare these with the size of the plots required to apply fertilizer treatments to the reclaimed land. In this situation the larger plots can be split into small plots on which the fertilizer treatments can be applied. This is a sensible design strategy on which to investigate the requirement for fertilizer after various test land-reclamation methods.

Split-plot experiments can be best thought of as an experiment within an experiment. Each of the two experiments has its own well-defined experimental unit. The only additional complication is that modelling of the interaction between the main plot and split-plot treatments.

There is an analogy between the application of the treatments to split plots and the application of treatments to plots within blocks. As in that case, the number of split plots per plot does not have to be the same for all plots. The same allocation strategy which has been used when blocks do not all contain the same number of plots also applies for the allocation of treatments to split plots.

The layout of split-plot experiments can be thought of as a nesting of small plots within an experiment requiring large plots. This nesting process can be continued for as long as it seems expedient. However, as more uncontrolled variation is added with the application of each set of treatments in the hierarchy, it must be expected that the standard error of parameter estimates from the largest experimental units will be larger than those from the units higher up in the hierarchy. This is often a strong disincentive to use split-plot experiments and consequently many experimenters try to avoid them.

5.3 RANDOM ALLOCATION OF TREATMENTS TO PLOTS

Fisher (1925) recognized that replication provided a way of estimating the variance of uncontrolled variation in an experiment so long as the plots within a block are assigned treatments at random.

The least-squares method of fitting a model provides unbiased estimates of treatment effects and valid tests of significance only if the deviations used to estimate the standard errors of the treatment responses are not correlated, and have an average of zero in the long run. Even when fertility gradients are present, random allocation of treatments ensures that this is true. The requirement for a long-run average of zero and uncorrelated deviations sounds a very artificial and mathematical requirement, but it has real practical meaning. This is apparent when considering the situation where the variation does not average zero in the long run, and when the deviations are correlated. An example is putting all replicate observations of one treatment on one side of a glasshouse bench and all observations from a control on the other. The deviations from the true value of the treatment response will probably not average zero in this situation as there may be an effect of the position of the treatment plots in the glass house added to the deviations from the true value. This added effect will also induce correlation amongst the deviations. If a least-squares model fitting is now performed, the estimated response to the treatments will be biased by virtue of the position of the treatment plots. In addition, the standard errors will not have the properties required of them in order to perform valid statistical tests and calculate valid confidence intervals.

For a valid least-squares analysis to be performed without randomization, any chosen systematic allocation of treatments to plots is presumed to not coincide with a pattern in the uncontrolled variation. This is a very difficult proposition to uphold without any evidence and, although the presumption may be true, there is no way of defending it against detractors. In addition, if a surprising result is obtained the experimenter may begin to have doubts about the objectivity of the layout.

Experiments laid out using random allocation do not suffer from this disadvantage. It may be that by a remote chance a systematic pattern does correspond to the random allocation of treatments. Even if this unlikely event does occur at least the experimenter cannot be accused of rigging the results by choosing an advantageous layout, and the dangers are not as great as not randomizing.

A random method of allocation means that all the possible allocations of treatments to the plots have an equal probability of being chosen. In this context 'all possible allocations' excludes those allocations which do not abide by the restrictions imposed by the chosen design. The random allocation of treatments to sets of three similar mares has already been considered in Example 4.5.4. and is an example of random allocation conditional upon the constraints imposed by the design. In a block experiment random allocation is restricted to the within-block allocation of treatments already assigned to that particular block. The number of possible allocations to a block with k plots is $1 \times 2 \times 3 \times ... \times k = k!$ (k factorial). With a Latin square design, once treatments are allocated to a single row or column, the next row or column can only be assigned from the subset of possibilities which comply with the restrictions imposed by the properties of a Latin square. That is, each treatment occurs just once in each row and column. Random allocation of treatments in a Latin square is most easily accomplished by randomly rearranging first the rows and then the columns of a standard Latin square.

There is a subtle difficulty with random allocation which it is important to avoid. The decision to include a particular treatment unit in an experiment should not be conditional on the treatment which is to be applied to it. If it is, the results of the experiment can be seriously biased. Examples of conditional inclusion are numerous. It is well known in medical research that potential recipients of an experimental treatment may agree to enter an experiment only on condition that they receive a treatment of their choice. Similarly, a researcher investigating the response of anthelmintic drugs on farm animals may want to exclude animals in poor condition from treatment groups destined to be dosed with a favoured formulation. If entry into an experiment is conditional on which treatment is to be applied, the treatment unit should not be considered for inclusion in the experiment in the first place.

The mechanics of randomly allocating treatments to plots (or rows or columns to positions in a Latin square) will now be described. Although the principle of randomization is often accepted, the way in which randomization is properly achieved is poorly understood. The problem is simply stated. Let the number of plots within a block, or rows and columns within a Latin square, be k. The basic requirement is to decide which of the $k!$ possible allocations to use for a particular experiment.

One technique is to write down all the $k!$ possibilities and index them 1 to $k!$. The allocation to use is then selected by drawing a random integer between 1 and k!. There are a number of ways of obtaining a random integer between 1 and $k!$.

(a) Put all the numbers on tags in a bag, and after shaking the bag draw one out.
(b) Playing cards can be used to obtain random numbers. The deck of cards can be as large as can be shuffled effectively. It is a good idea when many decks of cards are being combined to ensure they all have the same appearance.
(c) Work systematically through a table of random numbers, using numbers in sequence as they are required. If a number between 1 and 9 is required, use a single column of figures. If a number between 1 and 99 is required, use a double column of figures and so on. Care is needed when using random number tables to continue through the table from start to finish, bearing in mind it might take many months or years to work through the whole table. It is important to avoid the mistake of starting in the same place each time a random number is required.
(d) Assign the value zero or one to the results of heads or tails respectively from the toss of a coin. Then calculate a number between 1 and the power of 2 greater than $k!$ by the formula

$$1 + 1(\text{1st result}) + 2(\text{2nd result}) + \ldots + 2^{n-1}(n\text{th result}),$$

where n is the smallest integer such that $2^n \geq k!$. If the number calculated is greater than $k!$ then discard that number and repeat the calculation.
(e) Random numbers can be generated using a computer. Some random number generators have been found to be wanting, but most respectable statistical packages use satisfactory generators. However,

care is still needed when generating random integers. Suppose random numbers between 0 and 1 are generated and multiplied by 64. Truncation will yield random numbers between 0 and 63. Rounding will also not be without problems as, in the above example, the probability of obtaining 0 or 64 is just half that of the other numbers.

Another way of proceeding is to use tables of random permutations. These are available in Cochran and Cox (1957). They are used in the same way as has been described for random numbers in (c) above. Suppose we have six treatments to be assigned to six plots within a block and we have tables of random permutations of the integers 1 to 9. Let us also suppose the first random permutation is 8,1,5,7,9,3,6,4 and 2. We omit 7,8 and 9 from the list and assign treatment 1 to plot 1, treatment 5 to plot 2 and so on till treatment 2 is assigned to plot 6. The place in the table of random permutations would be marked and the next permutation used as the starting point for the next block.

A further way is to put one tag in a bag for each treatment to be applied to plots within a block. After shaking, draw the tags from the bag, allocating the treatment indicated on the tag to the first plot, and so on, till the treatment indicated on the last tag is allocated to the last plot. Playing cards can be used in a similar way. The treatment indicated on the first card is allocated to the first plot, and so on, till the treatment indicated on the last card is allocated to the last plot.

However, for many applications the above procedures are excessively complicated. Many statistical computer packages will make a random allocation of treatments to experimental plots with the supply of a single random number.

In some experimental situations it is possible to even out the effect of the environmental differences over the whole of the experiment by judicious repositioning of experimental material from time to time during the experiment. As an example, consider the reallocation of positions within a glass house to movable plots containing growing experimental material. This practice carries with it some dangers. In the glass-house example it might be supposed that a row and column design would allow for environmental gradients in the glass house. However, by continually rerandomizing pot plants which constitute the experimental plots, it could be argued that the effect of environmental gradients, the rows and columns, would be averaged out over each treatment plot. There is a danger in this strategy, however, as its adoption assumes there is no interaction between time (stages of growth) and position within the glass house. If there were, there would be a large treatment by replicate interaction which would not be allowed for in the analysis. This would inflate the estimates of variability, resulting in less powerful significance tests and inflated estimates of standard errors.

5.4 RANDOM SAMPLES FROM WITHIN PLOTS

It is common practice to make measurements on only a small part of a plot. In order to avoid observer bias it is necessary that the subplot from which the sample is taken is chosen at random. 'Random' is often interpreted to mean 'haphazard'. However, in practice haphazard is often not random, and this is easy to demonstrate. Yates (1935) provides some early examples. In one of these, some eight wheat plants were chosen, six at random and two haphazardly. When the experiment was done before the ears of wheat had formed there was a striking bias towards the taller plants with the haphazard sampling. After the ears had formed there was a strong bias towards the mid-sized plants.

A technique for obtaining a random sample from within a plot is described below. There is an exact parallel if sampling is to be from a group of animals. The area of the plot from which it has been decided to draw the sample is identified. This judgement may be based on the area of the plot which is expected to be free from interference from neighbouring plots. The identified part of the plot is now subdivided into subplots, each subplot being the sample size. The subplots are numbered and a random integer between 1 and the number of subplots is obtained by one of the techniques in section 5.3. The random integer identifies the subplot on which a measurement is to be made. If more than one subplot is to be measured, that number of random integers is drawn. The plots do not have to be physically subdivided into subplots in order to carry out the above procedure. A sheet of paper can be used to represent the field plot and the coordinates of the chosen subplot can be easily identified in the field plot.

This technique can be applied to each experimental plot to be harvested. However, it is common practice to harvest the subplot in the same position within each of the experimental plots. This random-start systematic sampling has two advantages. First, if there is a position effect within the plots it is eliminated by using subplots from the same position within each plot. Second, it is manageable, as a template can be made to identify easily the correct area of a plot to be harvested and so the possibility of error is reduced.

A similar technique involves randomly choosing the starting position on the border of a plot for a transect and then performing random-start systematic sampling along the transect. For example, a transect in the shape of a 'W' is often used for sampling soil within a plot. The starting position along the border of the plot is chosen at random and the 'W' is subdivided into the number of required sub-samples. The first subsample is taken at random from the first segment and subsequent subsamples are taken from the same place, or at random, in all subsequent segments.

A common technique is to throw a quadrat or other device into a plot and use this as an indicator of the area to sub-sample. This practice has a number of disadvantages. First, it is easy consciously or unconsciously to bias the results. This can be appreciated by spending a hot day throwing a quadrat and counting seedlings which lie within the boundary of the quadrat. It is hard to remain objective as the day progresses. It is also hard to accept a sample which does not comply with a preconceived judgement of what is representative of the plot. Many studies have been compromised by lack of attention to this detail.

Management of experiments and data 6

6.1 MANAGEMENT OF EXPERIMENTS

The problems encountered in managing experiments can be numerous, and many expensive experiments have been severely compromised and even ruined by lack of attention to this important aspect of experimentation. Most experimentalists have an experiment they would rather forget about. A catalogue of the experiments which I have been involved with and which suffer from this deficiency would be most embarrassing both to me and to my colleagues. Management problems may be attributed in part to the lack of dedication and discipline, but generally the major reason is insufficient skills to manage complex operations. The sorts of things which go wrong include:

- not starting the experiment on time;
- errors in laying out the experiment;
- haphazard servicing of the experiment;
- not making the specified measurements;
- not making the specified measurements on time;
- not making the specified measurements to the required standard;
- making measurements which were not required;
- unplanned bulking of samples from replicate plots before analysis.

Frequently such problems result from inadequate planning or from failure to follow a plan. These can be ameliorated by defining individual responsibilities, clearly specifying requirements and securing commitment from collaborators. The effort needed to guarantee freedom from management problems in the conduct of experiments, particularly multi-location and multi-manager experiments, should not be underestimated.

A planning calendar is a useful aid to ensuring that tasks are started and completed when planned. The use of such a tool which details when tasks are to be started, who is to perform them and important milestones such as the promised delivery date of essential supplies, provides a clear and comprehensible display of plans which are easily followed or reviewed. In complex situations, the use of formal project management tools should be considered. These began service in the late 1950s as tools for planning, scheduling, measuring and controlling development of military equipment. They have now become a part of managing many industrial processes. Their application when a complex experiment is being undertaken offers some considerable benefits as they assist in avoiding frequently encountered problems. Accounts of such methods can be found in many books, including Baker and Baker (1992) and Lowery (1992).

Although it is possible to use project management tools manually, there are many computer software implementations of the methodologies. It is also possible to use standard spreadsheet programs for some of the applications such as the preparation of Gantt tables and charts. Gantt tables and charts are used to detail the planned start and completion dates of the specific task identified in the translation of experimental intentions discussed in section 1.2. A simple example of a Gantt table and an incomplete chart is presented in Figure 6.1.1. The figure depicts the planned preparations for an imagined field test of a new seed drill. There are three separate strands in the preparations. The first involves the preparation of the drill, the second the arrangements for selecting and gaining access to the chosen test sites, and the third the supply and preparation of seeds and fertilizer. It should be noted that some tasks cannot be started before others have been completed, and these dependencies are noted in the 'Predecessors' column of the table and by the arrows linking bars on the chart. The dependencies result from physical constraints. For example, it is not possible to buy the seed and fertilizer before the requirements of the site have been assessed.

Note the insertion of a heavy line within the hatched bar indicating the second task listed on the chart. The line displays the progress which is being monitored daily because it is particularly important for the manager to ensure that progress is as planned. Also note the important milestones in task 8 which has been scheduled to be checked after four working days from the start of the project. This is to ensure the project manager is alerted to the need to make a critical decision concerning alternative arrangements should this task remain incomplete at the indicated time.

The tasks which are presented in the Gantt chart may then be subjected to network analysis. In the above example, one of these techniques, the critical path method (CPM), identifies the top path in the Gantt chart as being the critical path. That path is the longest path in

ID	Task Name	Duration	Start	Finish	Predecessors	Resource Names
1	**Prepare drill**	**45d**	**9/2/96**	**11/4/96**		
2	Finish drill design	8d	9/2/96	20/2/96		Elliot and Co.
3	Purchase components	10d	21/2/96	5/3/96	2	Purchasing Office
4	Assemble drill	25d	6/3/96	9/4/96	3	Workshop
5	Deliver drill	2d	10/4/96	11/4/96	4	Holgate and Co.
6	**Arrange sites**	**36d**	**9/2/96**	**29/3/96**		
7	Check on progress	0d	14/2/96	14/2/96		PDJ
8	Gain access to 12 sites	12d	9/2/96	26/2/96		SMP
9	Assess fertiliser status	20d	27/2/96	25/3/96	8	Soil lab
10	Assess seed required	5d	27/2/96	4/3/96	8	PDJ
11	Fence off areas	12d	27/2/96	13/3/96	8	SMP
12	Mark out exp. plan	12d	14/3/96	29/3/96	8, 11	SMP
13	**Obtain fertiliser and seed**	**12d**	**26/3/96**	**10/4/96**		
14	Purchase fertiliser and seed	10d	26/3/96	8/4/96	9, 10	Purchasing Office
15	Coat seed	2d	9/4/96	10/4/96	14	HP

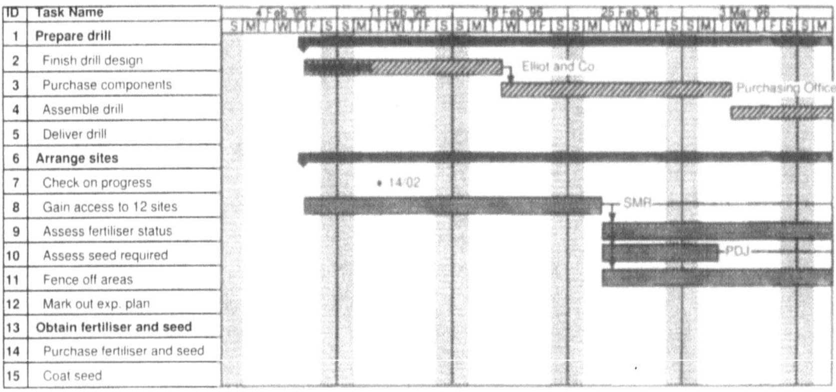

Figure 6.1.1 Gantt chart and an incomplete Gantt table.

terms of the time taken for its completion and consequently delays in tasks along this path would certainly delay the completion of the project. Reference to the chart enables sensible decisions to be made concerning the redeployment of resources should unplanned delays occur in the programme. For example, delays in the non-critical paths may be imposed by a manager who redeploys resources from them to make up for lost time in a critical path.

Figure 6.1.2 is the program evaluation and review technique (PERT) view of the same information. The boxes represent a completed stage in the project and the purpose of the chart is to represent relationships amongst the tasks.

The basic project management methodology can be made far more

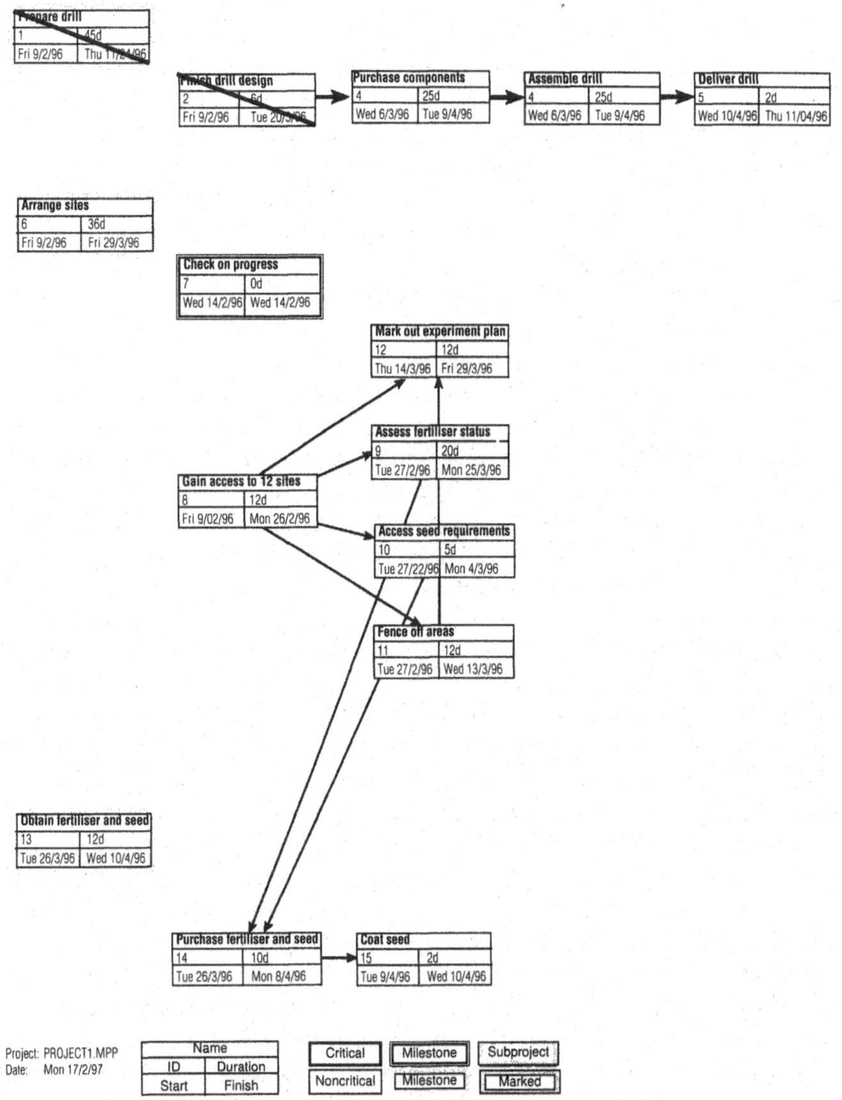

Figure 6.1.2 The PERT view of the information displayed in Figure 6.1.1.

sophisticated. For example, the time required for each activity can be a random variable and the project time can be simulated in order to establish the probability that a project will be finished by a specified date. This was the original use of PERT when the methodology was developed by Lockheed to assist its development of the Polaris missiles in the 1950s.

6.2 TOTAL QUALITY MANAGEMENT

Total quality management is another useful modern management tool. Application of the concept has had a large impact in improving productivity and profitability in the manufacturing industry. In modified form, it has also been applied to the service industry. It does not seem to have yet made a large impact on what might be called the research industry, where there is enormous potential for its profitable application.

There are many books, some written with missionary zeal, which encourage the adoption of total quality management. One account is given by Tenner and DeToro (1992).

In any organization, total quality management must develop from within and it should permeate the entire culture of an organization. It is motivated by an all pervading earnest commitment to obtain a better result by working through and with the people involved with the process. The necessary cultural changes take many years to develop. It took the Japanese manufacturing industry 20 years to surpass the quality and productivity of American and European companies. An approach in which a commitment today is a memory tomorrow is doomed to failure.

The central notion concerns a combination of manager commitment and participation to improved quality. In a successful implementation of the ideas, the organization is infused with the compulsion continually to improve processes with the full involvement of all workers.

Total quality management acknowledges and uses the fact that everyone enjoys being a visible and valued member of a successful team, and an essential ingredient is the will of the people to become involved. Total quality therefore requires the integration of team-building activities, good communication, encouragement with education, training and personal development, and above all recognition of achievements. Meetings of people involved with an experimental programme, which begin when the experiment is first conceived and continue till its completion, are an essential part of this process. Discussions which explain the purpose or progress of an investigation and include problem-solving sessions are a sure way of finding difficulties and their solutions. In many research organizations there is already a will to generate excellent results but the necessary culture has not been fostered.

There is a well-known plan, often called Shewhart's plan–do–check–act cycle, which is used to improve continuously a process which is part of the overall work programme. Although it now seems obvious, it is worthwhile writing it down and reflecting on it. It consists of the following steps:

- Plan specific actions which are intended to improve quality.
- Implement the plan.

- Check the results.
- If the results are acceptable, standardize on the improvements,

and so on. The cycle is operated continuously and the idea of a start and an end to the cycle is not admissible. Clearly the preparation of experimental plans would benefit from the application of this methodology. Similarly, improving the precision of the experiment by paying attention to the improvement of experimental technique would go some way to achieving long-term excellence.

The judge of the improvement in quality is the client. The client in this context is anyone, either internal or external to the organization, to whom a good or service is to be delivered. Therefore a central requirement is an understanding of what the client defines as good and bad quality. Good quality may well differ from the traditional view of high quality, the Rolls-Royces as it were, as clients make an unconscious compromise between this traditional view and the cost. It is important to recognize that part of perceived quality is the ability to deliver what is agreed, and on time. This is an important consideration in any research project.

6.3 DATA MANAGEMENT

Data management is the key to the final interpretation of experimental results. A careful disciplined approach is required to ensure that data are adequately documented and stored. A clear plan for managing the data as they are generated should be part of the experimental plan. In a complex multi-site experiment frustrations and time delays of years can be experienced in locating data because they have been stored in many forms and places. In this situation the solution to the problem is to consolidate the data storage at one site, perhaps in a relational data base. Specified subsets of the data can then be rapidly accessed for analysis and interpretation. In one example involving 19 sites and data from six years, it took six months for a persistent and persuasive worker to assemble all the data in a usable form at one location. Another important consequence of adopting the suggested strategy is to minimize the chances of losing the data.

The costs and effort required to achieve high-quality data management are very small when compared to the cost of collecting the data. However, many large experiments and some small experiments have never been fully analysed and interpreted because of lack of attention to this detail.

6.4 ACCURACY OF DATA RECORDING

The accuracy of data is important. It is paramount for experimental scientists to work on accurate data in order to maintain their scientific integrity, and to protect themselves from being easily discredited if their experiments become involved in litigation. The most sophisticated analysis will not overcome the potential problems caused by inaccurate data. There are a number of simple strategies which help to minimize the problem.

1. Use electronic data capture where possible. This avoids misreading and miswriting numbers. However, it may be necessary to convert the electronic data captured into a variate which is more meaningful. If, for example, it was important to know when a strong drying wind was blowing it might be better to convert wind run, wind speed, rainfall and temperature observations from a remote site to an index based on justifiable threshold values of those four variates.
2. Avoid manual transcription of data. The dangers of misreading and miswriting data can be easily demonstrated. A list of numbers transcribed by one member of a group and given to the next member for transcription and so on, soon results in considerable errors.

If data are to be manually recorded there are some important precautions which should be taken.

3. Collect data onto data recording sheets on which is written some form of unique identification. For crop experiments the unique identification could be the plot identification, which should include block and plot number within the block as well as treatment identification. For animal experiments the unique identification could be the tag number and colour as well as the treatment identification. This discipline imposes a constant checking procedure which helps ensure the data are entered in the correct position on the data sheet.
4. If feasible, the data recording sheets should have entries arranged in the order in which data are collected. The *n*th line on the data recording sheet should be for the *n*th observation. If data have to be entered in an unsystematic way onto a data recording sheet the chance of an entry being made in an incorrect place is high.
5. If data are entered manually into a computer they should be verified. This is best done by re-entering the data and reconciling disparities between the two computer files. A less successful method is for one person to call out the data from the computer file and another to check the values on the data recording sheets.
6. All data manipulations, including reordering, should be carried out on a computer. Data cannot be changed without outside interference once they are stored electronically. Additionally, electronically stored data can easily be stored securely to guard against loss from fire, theft or other catastrophe.
7. All calculations should be done by computer. Computers do not

make arithmetic mistakes or incorrectly read or copy a number once it has been entered.

8. Once data have been entered into the computer, required subsets should be obtained from the computer rather than by transcribing from the data recording sheets.
9. Document data files in an unambiguous way. After some time has passed it is easy to forget exactly what data a file contains, even if at the time of collection such a possibility seemed absurd.
10. Accurate manual data entry into a computer can be made more certain by including automatic range checking, checking for valid values when there are only a small number of possible values, and relational checks if it is known that one variate always has the same relation to another – for example, it might be always greater than or less than another.
11. Care is needed to ensure the precision of data is not compromised by poor calibration of instruments, reading instruments on too coarse a scale and parallax errors.

6.5 PRECISION OF DATA RECORDING AND ROUNDING

There is a need to consider other data recording issues. The number of significant digits in recording observations and rounding are important. Cochran and Cox (1957, p. 60) suggest:

> For the original data from which the analysis of variance is computed, a crude rule which errs on the safe side is as follows. Record the original data to 4 significant figures if the coefficient of variation of the experiment is 0.4% to 4.0%, to 3 if it is between 4% and 40% and 2 if it exceeds 40%.

The additional variance added to the uncontrolled variation in an experiment as a result of rounding can be calculated using the formula $d^2/12$, where d is the smallest possible difference between any two numbers after rounding. As an example, consider the situation when the last digit of a whole number is rounded to the nearest 10. The variance added to the uncontrolled variation is $10^2/12 = 8.3$.

In order to judge the importance of this additional source of variation, the square of the standard deviation of all other uncontrolled variation has this additional variance added to it and the square root of this new quantity is the new uncontrolled variation. The importance of the addition of this variance to the size of the other uncontrolled variation can then be judged for any particular situation.

In order to illustrate these ideas consider the consequence of rounding soil pH determinations to one decimal place. This is a common practice as more decimal places cannot be meaningfully interpreted. A typical coefficient of variation of pH determinations in experiments on soil is 1%, which for a soil with a pH of 5.5 means that the variance is $(0.01 \times 5.5)^2 = 0.003\ 025$. The component of this variance which results from rounding is $(0.1)^2/12 = 0.000\ 833$. Thus, in this situation, the practice of rounding to one decimal place results in generating 27.5% of the total experimental variation. The simple expedient of carrying two decimal places in pH measurements which are to be subjected to analysis would therefore substantially improve experimental precision.

Appendix A Catalogues and books containing experimental designs

Box, G.E.P. and Draper, N.R. (1987) *Empirical Model-building and Response Surfaces*. Wiley, New York.

Cochran, W.G. and Cox, G.M. (1957) *Experimental Designs*, 2nd edn. Wiley, New York.

Cornell, J.A. (1990) *Experiments with Mixtures: Designs, Models, and the Analysis of Mixture Data*. Wiley, New York.

Cox, D.R. (1958) *Planning of Experiments*. Wiley, New York.

Davies, O.L. (ed.) (1978) *The Design and Analysis of Industrial Experiments*. Longman Group, London for Imperial Chemical Industries.

Federer, W.T. (1993) *Statistical Design and Analysis for Intercropping Experiments*. Springer-Verlag, New York.

Ghosh, S. (1990) *Statistical Design and Analysis of Industrial Experiments*. Marcel Dekker, New York.

Gomez, K.A. and Gomez, A.A. (1976) *Statistical Procedures for Agricultural Research with Emphasis on Rice*. International Rice Research Institute, Manila.

John, J.A. and Williams, E. R. (1995) *Cyclic and Computer Generated Designs*. 2nd Edition. Chapman & Hall, London.

John, J.A. and Quenouille, M.H. (1977) *Experiments: Design and Analysis*. Macmillan, New York.

John, P.W.M. (1971) *Statistical Design and Analysis of Experiments*. Macmillan, New York.

Kempthorne, O. (1952) *The Design and Analysis of Experiments*. Wiley, New York.

Manly, B.F.J., McDonald, L.L. and Thomas, D.L. (1993) *Resource Selection by Animals: Statistical Design and Analysis for Field Studies*. Chapman & Hall, London.

Mead, R. (1988) *The Design of Experiments*. Cambridge University Press, Cambridge.

Mead, R., Curnow, R.N. and Hasted, A.M. (1993) *Statistical Methods in Agriculture and Experimental Biology*, 2nd edn. Chapman & Hall, London.

Montgomery, D.C. (1991) Design and Analysis of Experiments. Wiley, New York.

Appendix B
Computer programs to generate designs

ALPHA+
Biomathematics and Statistics Scotland
University of Edinburgh, Edinburgh EH9 3JZ, UK
http://www.bioss.sari.ac.uk/BioSS/computing.html

DSIGNX
Biomathematics and Statistics Scotland
University of Edinburgh, Edinburgh EH9 3JZ, UK
http://www.bioss.sari.ac.uk/BioSS/computing.html

GENSTAT
NAG Ltd, Wilkinson House,
Jordan Hill Road, Oxford OX2 8DR, UK
http://extweb.nag.co.uk/stats/TT.html

ECHIP
ECHIP, Inc.,
724 Yorklyn Rd., Hockessin, DE 19707, USA
http://www.echip.com/

Appendix C GENSTAT procedure POWER

PURPOSE

The procedure POWER calculates the power of a comparison for a specified real difference between two means, and variation between observations, with block effects removed if present, within treatment groups.

DESCRIPTION

The procedure allows for either completely randomized designs with equal replication of treatments, or a randomized block design in which each treatment occurs exactly once in each block. Any number of treatments in the design is allowed. When there are more than two treatments no allowance is made for adjusting the experimental error rate if multiple comparisons are planned.

From specified real differences in treatment means, and coefficient of variation, the probability of observing a significant result, using a two-tailed t test, is calculated by default for 2, 3, 4, 6, 8, 10, 15, 20, 25, 30, 40, 50, 75, 100, 150 and 500 replicates of the experiment.

The size of the significance test can be adjusted to any required probability.

Although the conditions under which the procedure is designed to operate may seem unduly restrictive, it is possible to use it in an imaginative way to cover other situations.

PARAMETERS

- DIFFERENCE is the size of the real difference to be detected.
- SD is the standard deviation of observations.
- DESIGN takes the value RB for randomized blocks or CR for completely randomized. The default is RB.

- NTREATS is the number of treatments in the design. The default is 2.
- REPS is the number of replicates of the treatments for which the power is calculated. The default is 2,3,4,6,8,10,15,20,25,30,40,50,75, 100,150, 500.
- ALPHA is the probability of rejecting the null hypothesis, given that it is true. The default is 0.05.

METHOD

The procedure uses the approximation to the non-central *t* distribution given in Abramowitz and Stegun (1970):

$$P(t' \mid \upsilon, \delta) \approx P(x)$$

where

$$P(x) = \frac{1}{\sqrt{2\pi}} \int_{-\infty}^{x} e^{-\frac{z^2}{2}} dz$$

and

$$x = \frac{t'(1 - \frac{1}{4\upsilon}) - \delta}{(1 + \frac{(t')^2}{2\upsilon})^{\frac{1}{2}}},$$

in which υ is degrees of freedom and δ is the non-centrality parameter.

REFERENCE

Abramowitz, M. and Stegun, I.A. (1970) *Handbook of Mathematical Functions*. Dover, New York.

EXAMPLE

POWER
DIFFERENCE=12.5;SD=7.5;DESIGN=CR;NTREATS=9;ALPHA=0.5

OUTPUT

Replication	Power
2	0.319
3	0.489
4	0.623
6	0.806
8	0.907
10	0.957

15	0.995
20	0.999
25	1.000
30	1.000
40	1.000
50	1.000
75	1.000
100	1.000
150	1.000
500	1.000

```
PROCEDURE 'POWER'

PARAMETER NAME= \
'DIFFERENCE', \
'SD', \
'DESIGN', \
'NTREATS', \
'REPS', \
'ALPHA'; \
MODE=2(p),t,3(p); \
NVALUES=4(1),*,1; \
VALUES=2(*),!T('RB','CR'),3(*); \
DEFAULT=2(*),'RB',2,!(2,3,4,6,8,10,15,20,25,30,40,50,75,100,150,500),.05; \
SET=yes; \
TYPE=!T(scalar),!T(scalar),!T(text),!T(scalar),!T(variate),!T(scalar); \
PRESENT-yes

EXIT [CONTROL=procedure; REPEAT=y; EXPLANATION='ALPHA is
out of bounds'] \
   (ALPHA.GE.1) .OR. (ALPHA.LE.0)

EXIT [CONTROL=procedure; REPEAT=y; EXPLANATION= \
   'NTREATS must be greater than 1'] (NTREATS.LE.1)
EXIT [CONTROL=procedure; REPEAT=y; EXPLANATION= \
   'NTREATS must be integer valued'] (NTREATS.NE.INT(NTREATS))

EXIT [CONTROL=procedure; REPEAT=y; EXPLANATION= \
   'MIN(REPS) must be greater than 1'] (MIN(REPS).LE.1)
EXIT [CONTROL=procedure; REPEAT=y; EXPLANATION= \
   'REPS must be integer valued'] MAX(REPS.NE.INT(REPS))

EXIT [CONTROL=procedure; REPEAT=y; EXPLANATION= \
   'DIFFERENCE must be positive'] (DIFFERENCE.LE.0)
```

```
EXIT [CONTROL=procedure; REPEAT=y; EXPLANATION= \
   'SD must be positive'] (SD.LE.0)

CALCULATE nr=NVALUES(REPS)
VARIATE [NVALUES=nr] df,tcrit,x[1,2],p[1,2],power,alpha1
SCALAR drb
CALCULATE drb=(DESIGN.EQS.'RB')
&       df=(NTREATS-drb)*(REPS-1)
&       alpha1=1-ALPHA
&       tcrit=SQRT(FED(alpha1;1;df))
&       x[1,2]=-1,1*tcrit*(1-1/(4*df))-SQRT(REPS/2)*DIFFERENCE/SD
&       p[1,2]=NORMAL(x[]/SQRT(1+tcrit**2/(2*df)))
&       power=1+p[1]-p[2]

PRINT
PRINT [IPRINT=*; SQUASH=yes] 'Replication','Power'; FIELD=12; DEC=0,3
PRINT [IPRINT=*; SQUASH=yes] REPS,power; dec=0,3
ENDPROCEDURE

RETURN
```

Glossary

Glossary of statistical terms used consistently throughout the text.

Accuracy

The closeness of an observation to the real value of the quantity being observed.

Bias

A systematic distortion in a statistic (q.v.).

Canonical variates

Used to predict the information in a set of variates (q.v.) from a second set of variates. This is done by finding a linear combination of the variates to be predicted and a linear combination of the predictors which has the largest correlation (q.v.). The two linear combinations are the canonical variates and the correlation between them is the canonical correlation. The second canonical variates are uncorrelated with the first and have the maximum correlation possible. In turn, each set of canonical variates has the maximum possible canonical correlation.

Categorical data

Data which are counts of observations which fall into well-defined groups. The groups may be descriptive or may have a natural order.

Confidence interval

An interval which is estimated by a statistic (q.v.), in which the population (q.v.) value of a parameter (q.v.) lies with a stated probability.

Correlation

A measure of the dependence between two variates. It is usually between –1 and 1, with a value of 0 indicating an absence of linear correlation.

Discriminant analysis

An application of canonical variates (q.v.).

Efficiency

An estimate of a parameter (q.v.) is more efficient than another if it has a smaller standard error.

Experimental unit

The experimental entity to which a single treatment is independently applied.

Factor

Used to indicate groupings of experimental units called levels.

Interaction

Interaction between factors (q.v.) is when the effect of levels of one factor depends on the levels of other factors.

Parameter

A variable quantity, occurring in a mathematical model of a particular phenomenon.

Population

The set of individuals about which an inference is made, usually by drawing a sample.

Power

The probability the observations from an experiment will reveal a real difference in treatment responses.

Precision

Refers to the closeness of an observation to the mean value of the dispersion of possible values. There is no suggestion that the mean value of the dispersion is the population value of which the observation is a member.

Principal components

Independent linear combinations of a set of variates (q.v.). In turn, each principal component has the maximum possible variation.

Significance test

A statistical technique used to detect values of test statistics (q.v.) which lie outside specified limits. If this occurs, the value is statistically significant.

Standard deviation

A measure of spread of a variate (q.v.).

Standard error

An estimate of the variation which will be encountered in estimates of parameters (q.v.) from repeated experiments on a population (q.v.).

Statistic

A value calculated from data.

Uncontrolled variation

Variation introduced into an observation by non-uniformity in experimental material, imprecision in treatment application and subsequent measurements.

Variable

A quantity which can take any one of a specified set of values.

Variance

A measure of the dispersion of a variate. It is the square of the standard deviation (q.v.).

Variance components

A partitioning of the variance into components which can be assigned to variation between the classes according to which the data are classified.

Variate

A random variable – that is, a quantity which can take any of the specified values with a specified relative frequency. Most experimental observations are on variates.

References

Altman, D. (1994) The scandal of poor medical research. *British Medical Journal* 308, 283–284

Anscombe, F.J. (1973) Graphs in statistical analysis. *American Statistician* 27, 17–21.

Baker, S. and Baker, K. (1992) *On Time – On Budget: A Step by Step Guide for Managing Any Project*. Prentice Hall, Englewood Cliffs, NJ.

Box, G.E.P. and Draper, N.R. (1987) *Empirical Model-Building and Response Surfaces*. Wiley, New York.

Box, G., Bisgaard, S. and Fung, C. (1988) An explanation and critique of Taguchi's contributions to quality engineering. *Quality and Reliability Engineering International* 4, 123–131.

Cochran, W.G. and Cox, G.M. (1957) *Experimental Designs*, 2nd edn. Wiley, New York.

Conniffe, D. (1976) A comparison of between and within herd variance in grazing experiments. *Irish Journal of Agricultural Research* 15, 39–46.

Cullis, B.R. and Gleeson, A.C. (1989) The efficiency of neighbour analysis for replicated variety trials in Australia. *Journal of Agricultural Science, Cambridge* 113, 233–239.

Cullis, B.R. and Gleeson, A.C. (1991) Spatial analysis of field experiments – an extension to two dimensions. *Biometrics* 47, 1449–1460.

Diggle, P.J. (1975) Robust density estimation using distance methods. *Biometrika* 62, 39–48.

Dobson, A. (1990) *An Introduction to Generalized Linear Models*. Chapman & Hall, London.

Draper, N. and Smith, H. (1966) *Applied Regression Analysis*, 2nd edn. Wiley, New York.

Fisher, R.A. (1925) *Statistical Methods for Research Workers*. Oliver and Boyd, Edinburgh.

Fisher, R.A. (1936) The use of multiple measurements in taxonomic problems. *Annals of Eugenics* 7, 179–188.

Gleeson, A.C. and Cullis, B.R. (1987) Residual maximum likelihood (REML) estimation of a neighbour model for field experiments. *Biometrics* 43, 277–288.

Hurlbert, S.H. (1984) Pseudoreplication and the design of ecological field experiments. *Ecological Monographs* 54, 187–211.

Jeffers, J.N.R. (1992) Role of biometry in environmental decision support. In *Proceedings of the XVI International Biometric Conference* (invited papers). Hamilton, NZ, pp. 137–147.

John, J.A. and Williams, E.R. (1995) *Cyclic and Computer Generated Designs*. Chapman & Hall, London.

Kempthorne, O. (1952) *The Design and Analysis of Experiments*. Wiley, New York.

Kenward, M.G. (1987) A method for comparing profiles of repeated measurements. *Applied Statistics* 36, 296–308.

Lowery, G. (1992) *Managing Projects with Microsoft Project*. Van Nostrand Reinhold, New York.

Lynch, P.B. (1966) *Conduct of Field Experiments*, Bulletin No. 399. New Zealand Department of Agriculture, Wellington.

Manly, B.F.J. (1986) *Multivariate Statistical Methods. A Primer*. Chapman & Hall, London.

McCance, I. (1995) Assessment of statistical procedures used in papers in the *Australian Veterinary Journal*. *Australian Veterinary Journal* 72(9), 322–328.

McCullagh, P. and Nelder, J.A. (1989) *Generalized Linear Models*, 2nd edn. Chapman & Hall, London.

Mead, R. (1988) *The Design of Experiments*. Cambridge University Press, Cambridge.

Mead, R. (1990) The non-orthogonal design of experiments. *Journal of the Royal Statistical Society A* 153, 151–201.

Miller, R.G. (1981) *Simultaneous Statistical Inference*, 2nd edn. Springer-Verlag, New York.

Papadakis, J.S. (1937) Méthode statistique pour des expériences sur champs. *Bulletin de l'Institute d'Amélioration des Plantes Salonique* 23.

Patterson, H.D. and Thompson, R. (1971) Recovery of inter-block information when block sizes are unequal. *Biometrika* 58, 545–554.

Pearson, E.S. and Hartley, H.O. (1970) *Biometrika Tables for Statisticians*, Volume 1. Cambridge University Press, Cambridge, for the Biometrika Trustees.

Popper, K. (1959) *The Logic of Scientific Discovery*. Hutchinson, London.

Ripley, B.D. (1987) *Stochastic Simulation*. Wiley, New York.

Robinson, D.L. (1987) Estimation and use of variance components. *The Statistician* 36, 3–14.

Seber, G.A.F. and Wild, C.J. (1988) *Nonlinear Regression*. Wiley, New York.

Silvey, S.D. (1975) *Statistical Inference*. Chapman & Hall, London.

Smith, H.F. (1938) An empirical law, describing heterogeneity in the yields of agricultural crops. *Journal of Agricultural Science* (Cambridge) 28, 1–23.

Stuart, A. and Ord, J.K. (1991) *Kendall's Advanced Theory of Statistics*, Volume 2. Oxford University Press, New York.

Tenner, A.R. and DeToro, I.J. (1992) *Total Quality Management*. Addison-Wesley, Reading, Mass.

Thompson, R. (1977) The estimation of heritability with unbalanced data. *Biometrics* 33, 485–495.

Underwood, A.J. (1994) On beyond BACI: sampling designs that might reliably detect environmental disturbances. *Ecological Applications* 4, 3–15.

Wilson, J.H. (1974) Discussion of D.J. Finney, 'Problems, data, and inference'. *Journal of the Royal Statistical Society A* 137, 20.

Yates, F. (1935) Some examples of biassed sampling. *Annals of Eugenics* 6, 202.
Yates, F. (1964) Sir Ronald Fisher and the design of experiments. *Biometrics* 20, 307–321.
Yule, G.U. (1926) Why do we sometimes get nonsense-correlations between time-series? A study in sampling and the nature of time series. *Journal of the Royal Statistical Society* 89, 1.

Index